AF413270

FUNDAMENTAL AND APPLIED PROBLEMS IN TERAHERTZ-RELATED DEVICES AND TECHNOLOGIES

FUNDAMENTAL AND APPLIED PROBLEMS IN TERAHERTZ-RELATED DEVICES AND TECHNOLOGIES

Editors

Taiichi Otsuji
Tohoku University, Japan

Wojciech Knap
Polish Academy of Sciences, Poland

Maxim Ryzhii
University of Aizu, Japan

Michael Shur
Rensselaer Polytechnic Institute, USA

World Scientific

NEW JERSEY · LONDON · SINGAPORE · BEIJING · SHANGHAI · TAIPEI · CHENNAI

Published by

World Scientific Publishing Co. Pte. Ltd.

5 Toh Tuck Link, Singapore 596224

USA office: 27 Warren Street, Suite 401-402, Hackensack, NJ 07601

UK office: 57 Shelton Street, Covent Garden, London WC2H 9HE

Library of Congress Control Number: 2025007200

British Library Cataloguing-in-Publication Data
A catalogue record for this book is available from the British Library.

Selected Topics in Electronics and Systems — Vol. 69
FUNDAMENTAL AND APPLIED PROBLEMS IN
TERAHERTZ-RELATED DEVICES AND TECHNOLOGIES

Copyright © 2025 by World Scientific Publishing Co. Pte. Ltd.

ISBN 978-981-12-8680-3 (hardcover)
ISBN 978-981-12-8681-0 (ebook for institutions)
ISBN 978-981-12-8682-7 (ebook for individuals)

For any available supplementary material, please visit
https://www.worldscientific.com/worldscibooks/10.1142/13691#t=suppl

Desk Editors: Soundararajan Raghuraman/Steven Patt

Typeset by Stallion Press
Email: enquiries@stallionpress.com

Contents

Chapter 1

Plasmonic Crystals for Terahertz Detection, Amplification, and Generation[#]

Michael Shur[*,†,§], **John Mikalopas**[‡,¶], **and Gregory Aizin**[‡,||]

†*Department of Electrical, Computer and Systems Engineering*
Rensselaer Polytechnic Institute
110 8th Street, Troy, NY 12180, USA
‡*Kingsborough College, The City University of New York*
Brooklyn, NY 11235, USA
§*shurm@rpi.edu*
¶*jmikalopas@kbcc.cuny.edu.edu*
||*gaizin@kbcc.cuny.edu*

TeraFET arrays operating in plasmonic regimes could support the transition from 5G to 6G communication if the constituent TeraFETs operate in synchrony. Such arrays are plasmonic crystals supporting Bloch-like waves of electron density oscillations. The key issues are breaking symmetry and maintaining appropriate boundary conditions between the unit cells. The symmetry must be broken to choose the response polarity to detect the direction of the plasmonic instability

---

*Corresponding author.

[#]This chapter appeared previously on the International Journal of High Speed Electronics and Systems. To cite this chapter, please cite the original article as the following: M. Shur, J. Mikalopas and G. Aizin, *Int. J. High Speed Electron. Syst.*, **33**, 2540026 (2024), doi:10.1142/S0129156425400269.

growth for generating THz oscillations. The coherence of plasma waves propagating in individual cells of the plasmonic crystal results in continuous waves in the entire structure. Using the narrow stripes at the unit cell edges (called plasmonic stubs) could maintain such coherence. Another advantage of TeraFET arrays is the reduced effects of parasitic contact resistance. This advantage is even more pronounced in ring plasmonic structures used for converting THz radiation into a magnetic field (giant inverse Faraday effect).

Keywords: THz; FETs; SPICE; compact model; TeraFET; plasmonic crystal.

1. Introduction

Short channel Field Effect Transistors could operate as TeraFETs detecting [1–5] and even emitting sub-THz and THz radiation [6–10]. These devices have the potential to enable 6G communication in the 300 GHz band [11–14]. As shown in Fig. 1, the transition of 5G to 6G will become unavoidable around 2028 to support an ever-growing data traffic. Such a transition requires better THz detectors and emitters and could be supported by TeraFET arrays operating in plasmonic regimes [15–20] with the constituent TeraFETs working in synchrony.

Hydrodynamic equations describing plasma waves are similar to those describing water or sound waves [21–23]. Figure 2

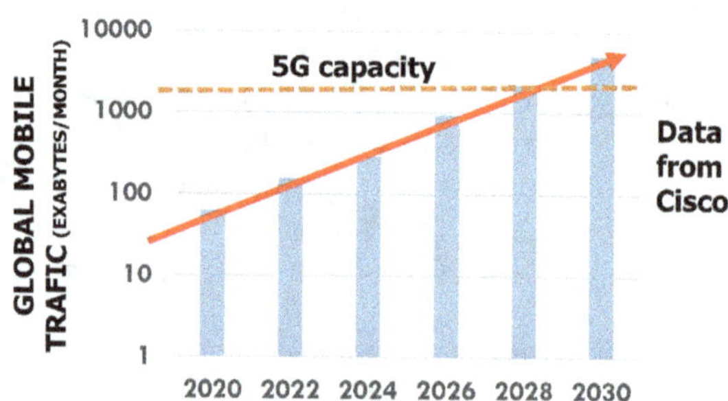

Fig. 1.   Global mobile data traffic in exabytes/month. The dotted line shows an expected limit for 5G.

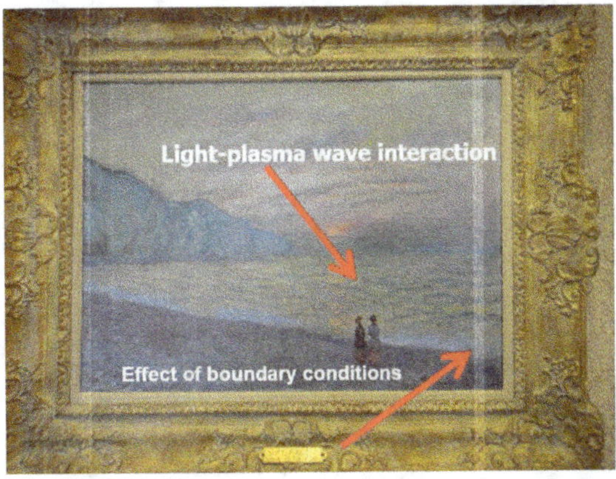

Fig. 2. Claude Monet Impressions of Sunset, Pourville (1882), Kreeger Museum, Washington DC.

Fig. 3. Whirlpool (Utagawa Hiroshige (1797–1858) "Naruto Whirlpool, Awa Province ..." (The Metropolitan Museum of Art, New York, NY) and twisted plasmons (after [24]).

illustrates the comparison between plasma waves and water waves. The physical effects displayed in this painting are similar to those that might be observed in the electronic fluid, such as light-plasma wave interaction or the effects related to the boundary conditions. Figures 3 and 4 further illustrate this analogy which applies to

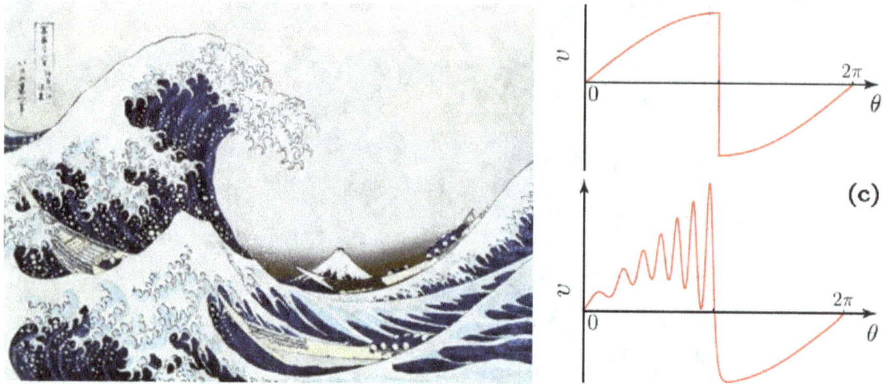

Fig. 4. Shock water wave (Hokusai 1760–1849) and shock plasma wave pattern: electron fluid velocity in a ring versus the angular coordinate (from [25]).

the exciting phenomena happening in water at high excitation levels.

2. Instability Mechanisms in Plasmonic Crystals

Plasma waves in plasmonic crystals might become unstable. Their growth leads to electromagnetic radiation at sub-THz or THz frequencies. The instability mechanisms include "shallow water" instability of the plasma waves in a ballistic FET due to the plasma wave excitations by a DC current and plasma wave reflections from the channel boundaries (see Fig. 5) [15, 26–28], transit time instability [20], and plasmonic boom instability [29–32].

The idea of transit time instability is to achieve the 180° phase shift between the waveforms of the AC current and AC voltage. This is achieved by having a transit time delay of the electrons traveling across the region between the gated channel and the drain contact to shift the phase of the current related to the changes of electron density in the channel induced by the plasma oscillations with respect to the voltage variation by 180° (see Fig. 6).

Figure 7 illustrates the concept of a "plasmonic boom" device [29–31]. The idea is to create sections of the device to ensure the

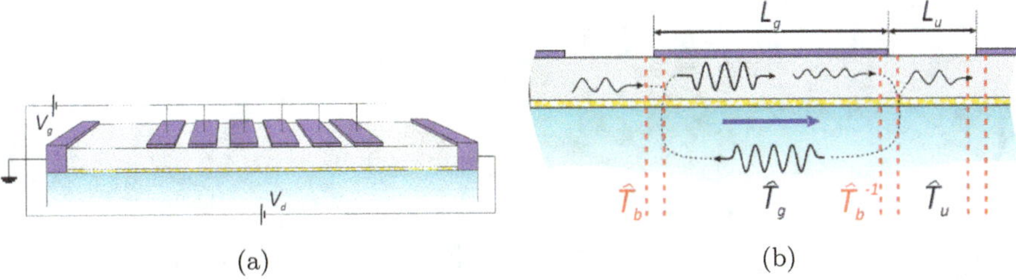

(a)   (b)

Fig. 5.   Multifinger gate TeraFET structure (a) and superposition of plasma waves (b) (from [15]).

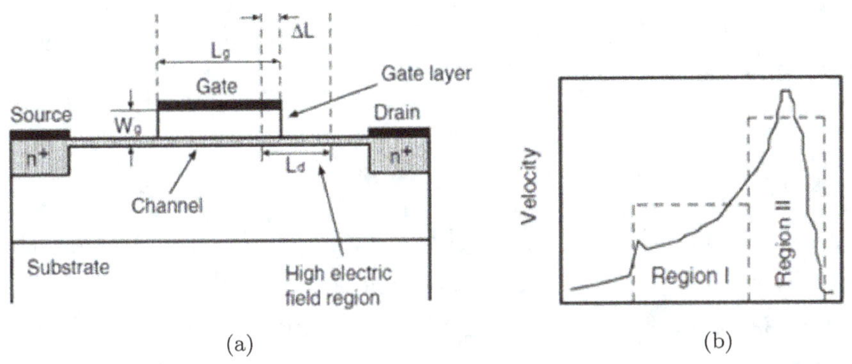

(a)   (b)

Fig. 6.   Velocity distribution in a short channel FET the electron transit time across Region 2 creates the phase shift (from [20]).

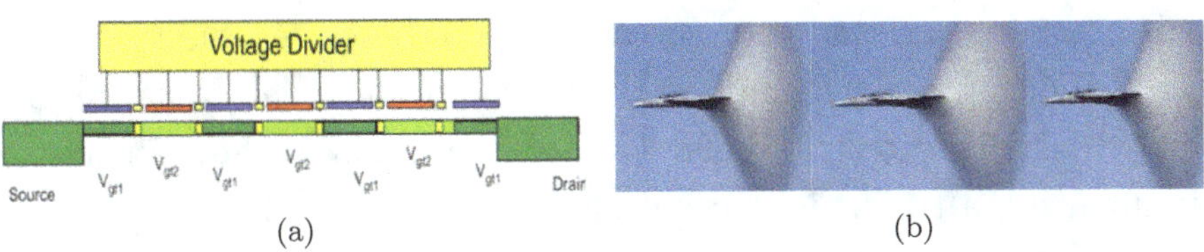

(a)   (b)

Fig. 7.   Schematic of "plasmonic boom" device (a) [29] and sound boom analogy (b).

electrons transfer from one section to the next with the electron drift velocity crossing the plasma velocity threshold so that velocity changes from being smaller than the plasma velocity to being larger than the plasma velocity. This is similar to the sonic boom illustrated in Fig. 7(b) but in the plasmonic boom structure, such

                    M. Shur et al.

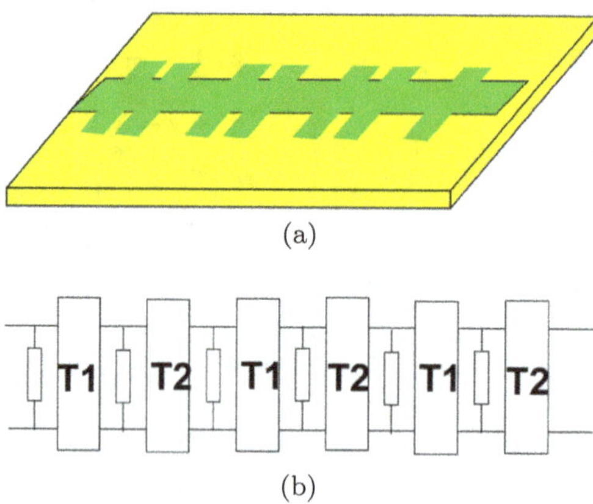

(a)

(b)

Fig. 8.   Two-dimensional electron channels of variable width (a) and the equivalent circuit (b) (after [32]).

a transition could occur many thousands of times. As a result, the generated THz power could reach hundreds of milliwatts [32]. The first observations of the plasmonic boom instability have already been reported [30].

The key issue in achieving a synchronous operation of all the units of a plasmonic crystal is maintaining the boundary conditions and adjusting the phases of the plasma waves at the boundaries. This could be achieved using a variable width design illustrated by Fig. 8 showing the streamlines of the electron current deviating in the extension regions at the section boundaries and the equivalent circuit using a transmission line model.

This model predicts that plasma waves in a plasmonic crystal have a dispersion law similar to the Bloch waves in a conventional crystal as shown in Fig. 9. This figure shows the frequencies (f) and increments/decrements (w'') of the plasma modes excited by an external EM wave in the graphene dual-grating-gate TeraFET [33] as a function of the drift velocity v_{02} proportional to the drain voltage. As seen, some modes crossover as the drain voltage

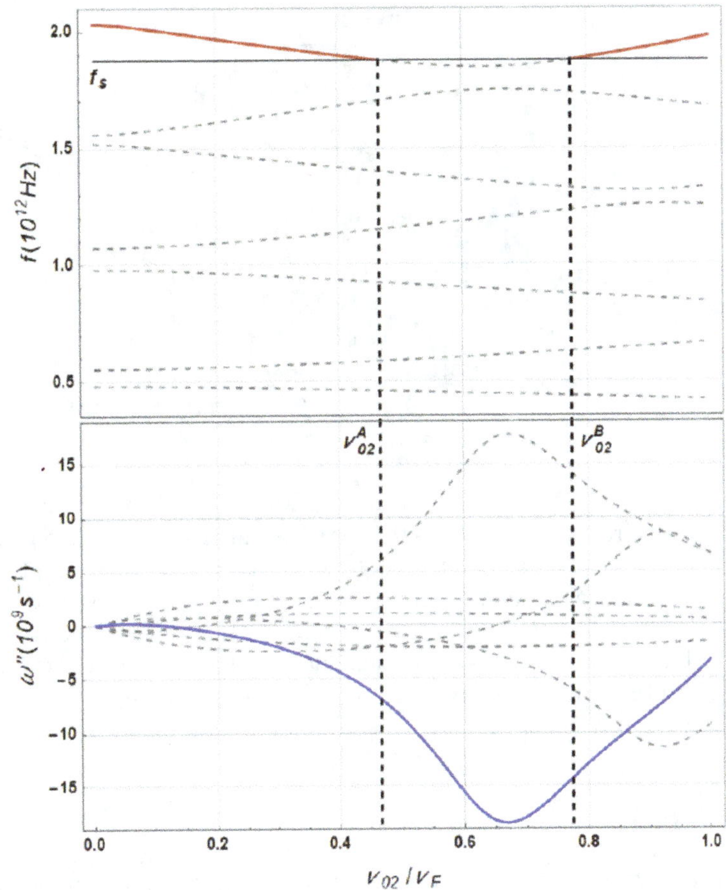

Fig. 9.   Frequencies (f) and increments/decrements (w'') of the plasma modes excited by an external EM wave in the graphene dual-grating-gate TeraFET as a function of drift velocity v_{02} (proportional to the drain voltage). Both the damped modes (dashed grey lines) and the active mode (red and blue lines) are shown; f_S is the threshold mode frequency with the quality factor $Q = w\tau = 1$; are the drift velocities at the boundaries (dashed vertical lines) of the gap in the detected plasmonic spectrum [33].

increases switching from damped to active modes (the Amplified Mode Switching (AMS) effect) that explains the red and blue shifts of the observed plasmonic peaks and switching from attenuation to transparency and amplification with the drain bias (see Fig. 10).

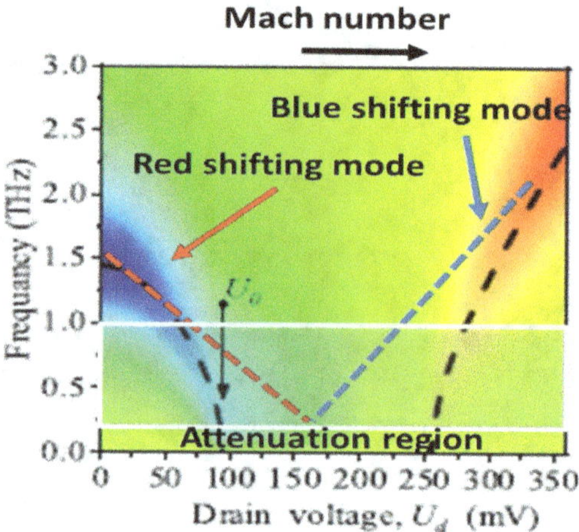

Fig. 10. Illustration of Amplified Switching Mode (AMS) effect observed in Ref. 34. The blue region corresponds to the redshifting mode. In this region, the attenuation of the THz radiation was observed. The semi-transparent region corresponds to the situation, where the frequency is becoming too low in comparison with the inverse decrement to excite the plasmonic peak. In the red region, the blue-shifted mode is emerging, and its increment is large enough to decrease the THz absorption as observed in Ref. 34 (from [35]).

3. Conclusion

Plasmonic crystals have the potential to improve the performance of TeraFET detectors of THz radiation and provide efficient and compact electronic sources of sub-THz and THz radiation because they could combine nano sizes of individual cells with macro sizes of the entire crystal. Using the variable width geometry combining the narrow protruding plasmonic regions — "plasmonic stubs" and wider longer cell sections provides the proper boundary conditions between the cells enabling the plasmonic instabilities. Bloch-like dispersion of plasma waves in plasmonic crystals supports the crossover of active and decaying mode and explains recent observations of the AMS effect in graphene plasmonic crystals with dual grating.

Acknowledgment

The work was supported by ARO (Program Manager Dr. Joe Qiu).

ORCID

Michael Shur https://orcid.org/0000-0003-0976-6232

John Mikalopas https://orcid.org/0000-0002-9946-8716

Gregory Aizinz https://orcid.org/0000-0002-0017-3417

References

1. M. Dyakonov and M. Shur, "Detection, mixing, and frequency multiplication of terahertz radiation by two-dimensional electronic fluid," IEEE Electron Dev, vol. 43, No. 3, pp. 380–387, 1996.
2. W. Knap *et al.*, "Nonresonant detection of terahertz radiation in field effect transistors," J Appl Phys, vol. 91, No. 11, pp. 9346–9353, 2002.
3. W. Stillman, M. Shur, D. Veksler, S. Rumyantsev, and F. Guarin, "Device loading effects on nonresonant detection of terahertz radiation by silicon MOSFETs," Electron Lett, vol. 43, No. 7, pp. 422–423, 2007.
4. D. Veksler, F. Teppe, A. Dmitriev, V. Y. Kachorovskii, W. Knap, and M. Shur, "Detection of terahertz radiation in gated two-dimensional structures governed by DC current," Phys Rev B, vol. 73, No. 12, p. 125328, 2006.
5. W. Knap, F. Teppe, Y. Meziani, N. Dyakonova, J. Lusakowski, F. Bouef, T. Skotnicki, D. Maude, S. Rumyantsev, and M. S. Shur, "Plasma wave detection of millimeter wave radiation by silicon field effect transistors," Appl Phys Lett, vol. 85, No. 4, pp. 675–677, 2004.
6. M. Dyakonov and M. Shur, "Shallow water analogy for a ballistic field effect transistor: New mechanism of plasma wave generation by dc current," Phys Rev Lett, vol. 71, No. 15, p. 2465, 1993.
7. T. Otsuji *et al.*, "Emission and detection of terahertz radiation using two-dimensional electrons in III–V semiconductors and graphene," IEEE Trans Terahertz Sci Technol, vol. 3, No. 1, pp. 63–71, 2013, doi:10.1109/TTHZ.2012.2235911.
8. W. Knap, D. Coquillat, N. Dyakonova, F. Teppe, O. Klimenko, H. Videlier, S. Nadar, J. Lusakowski, G. Valusis, F. Schuster, B. Giffard, T. Skotnicki, C. Gaquière, and A. El Fatimy, "Plasma excitations in field effect transistors for terahertz detection and emission," CR Phys, vol. 11, No. 7–8, pp. 433–443, 2010.

9. V. Ryzhii, T. Otsuji, and M. Shur, "Graphene plasmonics for terahertz applications," Appl Phys Lett, vol. 116, 140501, 2020, doi:10.1063/1.5140712.

10. W. Knap, J. Lusakowski, T. Parenty, S. Bollaert, A. Cappy, V. Popov, and M. S. Shur, "Emission of terahertz radiation by plasma waves in 60 nm AlInAs/InGaAs high electron mobility transistors," Appl Phys Lett, vol. 84, No. 13, pp. 2331–2333, 2004.

11. M. Shur, G. Aizin, T. Otsuji, and V. Ryzhii, Plasmonic Field-Effect Transistors (TeraFETs) for 6G communications, Sensors, vol. 21, No. 23, p. 7907, 2021.

12. C. Jastrow, K. Munter, R. Piesiewicz, T. Kurner, M. Koch, and T. Kleine-Ostmann, "300 GHz transmission system," Electron Lett, vol. 44, pp. 213–214, 2008.

13. H. J. Song, K. Ajito, A. Hirata, A. Wakatsuki, Y. Muramoto, T. Furuta, N. Kukutsu, T. Nagatsuma, and Y. Kado, "8 Gbit/s wireless data transmission at 250 GHz," Electron Lett, vol. 45, pp. 1121–1122, 2009.

14. M. Shur, "Plasmonic detectors and sources for THz communication and sensing," SPIE Proc, vol. 10639, Micro- and Nanotechnology Sensors, Systems, and Applications X, 1063929 (2018).

15. A. S. Petrov, D. Svintsov, V. Ryzhii, and M. S. Shur, "Amplified-reflection plasmon instabilities in grating-gate plasmonic crystals," Phys Rev B, vol. 95, p. 045405, 2017.

16. I. V. Gorbenko, V. Kachorovskii, and M. Shur, "Terahertz plasmonic detector controlled by phase asymmetry," Optics Express, vol. 27, No. 4, 2019, arXiv:1901.02036.

17. M. S. Shur, X. Liu, S. Rumyantsev, and V. Kachorovskii, "Heterodyne phase sensitive terahertz spectrometer," in Proc 2017 IEEE Sensors Conference, pp. 621–623.

18. M. Shur, "Silicon and silicon-germanium terahertz electronics," in Proc IEEE MTT-S International Microwave Workshop Series on Advanced Materials and Processes for RF and THz Applications, (IMWS-AMP 2018), July 16–18, 2018, Ann Arbor, MI, USA, 978-1-5386-5569-6/18/$31.00.

19. M. Shur, S. Rudin, G. Rupper, and T. Ivanov, "p-Diamond as candidate for plasmonic terahertz and far infrared applications," Appl Phys Lett, vol. 113, p. 253502, 2018, doi:10.1063/1.505309.

20. V. Ryzhii, A. Satou, and M. S. Shur, "Transit-time mechanism of plasma instability in high-electron mobility transistors," Phys Stat Sol A, vol. 202, No. 10, pp. R113–R115, 2005.

21. M. I. Dyakonov and M. S. Shur, New Hydrodynamic Phenomena in Electronic Fluid, in "22nd Int Conf Physics of Semiconductors," D. J. Lockwood, Editor, World Scientific, vol. 1, pp. 811–814, 1995.

22. S. Rudin, G. Rupper, and M. Shur, "Ultimate response time of high electron mobility transistors", J Appl Phys, vol. 117, p. 174502, 2015, doi:10.1063/1.4919706.

23. Y. Zhang and M. S. Shur, "p-Diamond, Si, GaN, and InGaAs TeraFETs," IEEE Trans Electron Dev, vol. 67, No. 11, pp. 4858–4865, 2020, doi:10.1109/TED.2020.3027530.
24. S. O. Potashin, V. Yu. Kachorovskii, and M. Shur, "Hydrodynamic inverse faraday effect in two-dimensional electron liquid," Phys Rev B, vol. 102, p. 085402, 2020, arXiv:2001.08015v1 [cond-mat.mes-hall].
25. K. L. Koshelev, V. Y. Kachorovskii, M. Titov, and M. S. Shur, "Plasmonic shock waves and solitons in a nanoring," Phys Rev B, vol. 95, p. 035418, 2017.
26. T. Otsuji *et al.*, "Emission and detection of terahertz radiation using two-dimensional electrons in III–V semiconductors and graphene," IEEE Trans Terahertz Sci Technol, vol. 3.1, 63–71, 2013.
27. T. Watanabe, A. Satou, T. Suemitsu, W. Knap. V. V. Popov, and T. Otsuji, "Plasmonic terahertz monochromatic coherent emission from an asymmetric chirped dual-grating-gate InP-HEMT with a photonic vertical cavities," Conf Lasers and Electrooptics Dig., CW3K.7, San Jose, CA, USA, June 12, 2013.
28. G. R. Aizin, J. Mikalopas, and M. Shur, "Current driven Dyakonov–Shur instability in ballistic nanostructures with a stub," Phys Rev Appl, vol. 10, p. 064018, 2018.
29. V. Yu. Kachorovskii and M. S. Shur, "Current-induced terahertz oscillations in plasmonic crystal," Appl Phys Lett, vol. 100, p. 232108, 2012.
30. T. Hosotani, A. Satou, and T. Otsuji, "THz emission in a dual-grating-gate HEMT promoted by the plasmonic boom instability," 2021 46th Int Conf Infrared, Millimeter and Terahertz Waves (IRMMW-THz), Chengdu, China, 2021, pp. 1–2, doi:10.1109/IRMMW-THz50926.2021.9567205.
31. G. R. Aizin, J. Mikalopas, and M. Shur, "Current driven "plasmonic boom" instability in gated periodic ballistic nanostructures," Phys Rev B, vol. 93, No. 19, 195315, 2016.
32. G. R. Aizin, J. Mikalopas, and M. Shur, "Plasmonic instabilities in two-dimensional electron channels of variable width," Phys Rev B, vol. 101, p. 245404, 2020.
33. G. Aizin, J. Mikalopas, and M. S. Shur, "Plasma instability and amplified mode switching effect in THz field effect transistors with grating gate," Phys Rev B, vol. 107, p. 245424, 2023, https://arxiv.org/abs/2303.16152.
34. S. Boubanga-Tombet, W. Knap, D. Yadav, A. Satou, D. B. But, V. V. Popov, I. V. Gorbenko, V. Kachorovskii, and T. Otsuji, "Room-temperature amplification of terahertz radiation by grating-gate graphene structures," Phys Rev X, vol. 10, p. 031004, 2020.
35. M. Shur, J. Mikalopas, and G. Aizin, "Amplified mode switching effect in THz field effect transistors with grating gate," 2023 48th Int Conf Infrared, Millimeter, and Terahertz Waves (IRMMW-THz), Montreal, QC, Canada, 2023, pp. 1–2, doi:10.1109/IRMMW-THz57677.2023.10299347.

Chapter 2

Terahertz Wave Power Multiplication by Combining Photocurrents from Arrayed UTC-PDs[#]

**Hussein Ssali[*], Yoshiki Kamiura[†],
Ryo Doi[‡], Hiroki Agemori[§],
Ming Che[¶], Yuya Mikami[‖], and
Kazutoshi Kato[**]**

*Graduate School of Information Science
and Electrical Engineering, Kyushu University,
744 Motooka, Nishi-ku, Fukuoka, Japan*
[*]ssali.hussein.556@s.kyushu-u.ac.jp
[†]kamiura.yoshiki.639@s.kyushu-u.ac.jp
[‡]doi.ryo.206@s.kyushu-u.ac.jp
[§]agemori.hiroki.551@s.kyushu-u.ac.jp
[¶]che.ming.677@m.kyushu-u.ac.jp
[‖]mikami@ed.kyushu-u.ac.jp
[**]kato@ed.kyushu-u.ac.jp

[*]Corresponding author.
[#]This chapter appeared previously on the International Journal of High Speed Electronics and Systems. To cite this chapter, please cite the original article as the following: H. Ssali, Y. Kamiura, R. Doi, H. Agemori, M. Che, Y. Mikami and K. Kato, *Int. J. High Speed Electron. Syst.*, **33**, 2440025 (2024), doi:10.1142/S0129156424400251.

A terahertz (THz) wave generator for multiplying the output power from arrayed photomixers is developed. The generator is composed of two uni-traveling carrier photodiodes (UTC-PDs) as the photomixers, a 2×1 T-junction power combiner and a microstrip rectangular patch antenna on a silicon carbide (SiC) substrate in which photocurrents of 300 GHz from the UTC-PDs are combined to be fed to the antenna. Experimental results show that the output power is multiplied by 5.8 dB (approximately fourfold) compared to that of a single UTC-PD. This advancement has the potential for further multiplication of THz power by cascaded combining of a larger number of UTC-PDs.

Keywords: Microstrip line; patch antenna; terahertz wave; T-junction power combiner; uni-traveling carrier photodiode (UTC-PD).

1. Introduction

The rapid growth of communication applications, such as e-commerce, cloud computing, video conferencing and collaboration tools, coupled to the growing global connectivity has led to a significant increase in data traffic worldwide. One of the candidates to address the demand for ultra-high data rates is the terahertz (THz) wave technology. THz waves (0.1–10 THz), with their unique features such as wide bandwidth and penetration capabilities, have potential applications not only in wireless communication but also in sensing and imaging [1, 2]. Due to its distinctive position within the electromagnetic spectrum, researchers in the fields of electronics and photonics have come together to address the technological challenges in this area. As a result of their collective efforts, various technologies for generating THz signals have made significant progress over the past decade. In the realm of electronics, III–V-based semiconductor technologies, such as InP heterojunction bipolar transistors (HBTs), high electron mobility transistors (HEMTs) and GaAs-based Schottky diodes, have demonstrated the capability to generate power levels in the range of $100 \, \mu\text{W}$–100 mW at room temperature [3–7]. Photonics methods for THz generation have also undergone significant developments. Photomixer-based approaches especially using a uni-traveling carrier photodiode (UTC-PD) have achieved tens of μW of continuous-wave power [8–11]. The salient challenge with both electronics- and photonics-based sources is the drastic output power decrease as frequencies get higher [12].

Current combining from arrayed UTC-PDs is one of the feasible approaches to increase the output power beyond what a single UTC-PD can provide [13]. The implementation of microstrip line (MSL) transmission technology is a suitable method for large-scale current combining because the geometry is easier to be cascaded than other types of lines such as the coplanar waveguide (CPW). Despite the intricacy of a large-scale combiner, it is important to note that it essentially consists of 2×1 combiners. In our previous work, we generated a 300-GHz wave by the monolithic integration of two UTC-PDs with an on-chip Wilkinson combiner and a patch antenna, as referenced in [14, 15]. The combiner has a resistor between two branches which would cause a power loss with unbalanced photocurrents.

In this study, we examine the performance of photocurrent combination by a 2×1 T-junction combiner that does not have a resistor. Though the T-junction provides lower isolation between the two branches, the unnecessity of a resistor as well as the simple layout enable us to reduce combining losses, which would likely have a significant impact on the combining efficiency at such high frequencies. We present the experimental characterization of an integrated device in terms of radiated power with respect to the photocurrent from UTC-PD. Because the UTC-PD is a current source, the current delivered to the antenna is ideally doubled, resulting in a fourfold increase in radiated power. In addition, a silicon carbide (SiC) substrate is applied to the integrated device because of its high thermal conductivity [16–19] which prevents thermal damage to the UTC-PD at a high photocurrent.

2. THz Wave Generation by Photomixing with a UTC-PD

Photomixing with a UTC-PD is a technique that enables the conversion of two lightwaves with different frequencies into an alternating current with a frequency equal to the difference between the two lightwaves [20, 21]. This method is regarded as a promising approach for generating terahertz waves, especially in wireless communications, imaging and sensing applications. The uni-traveling carrier photodiode (UTC-PD) was developed by NTT Laboratories for high-speed optical communication systems [22]. Since its invention, extensive research has been conducted to enhance its capabilities and explore potential applications [23, 24]. The UTC-PD offers the unique advantage of simultaneously achieving ultra-high speed and high output power, surpassing the conventional pin-PD. This exceptional characteristic renders it highly suitable for electromagnetic (EM)-wave generation. Figure 1 illustrates the UTC-PD-based photomixing technique. By combining two lightwaves with different frequencies (f_1 and f_2), a

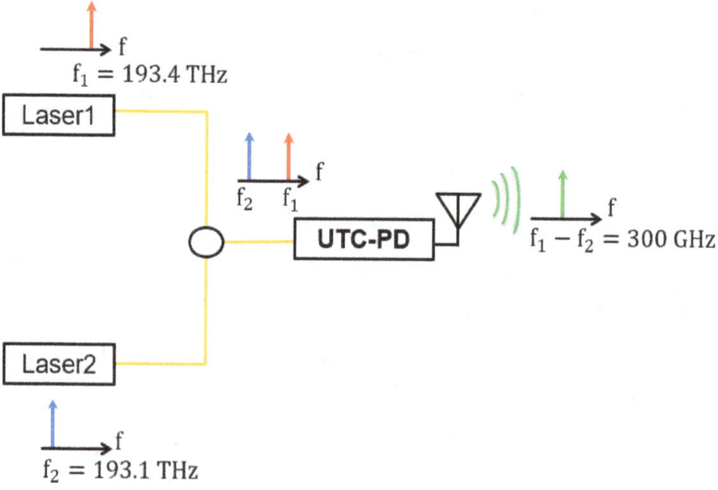

Fig. 1.   UTC-PD-based photomixing for THz wave generation.

beat-signal with a beating frequency (f_b) equal to the difference of f_2 and f_1 is obtained after photomixing. For instance, a 1550-nm-band lightwave with frequencies of 193.4 THz and 193.1 THz is converted into an alternating electric current with the same frequency as f_b when input into the UTC-PD. THz waves can be generated by the modulated current through an antenna or waveguide.

3.   Power Combining with a T-Junction Combiner

Multiplication of THz wave power by combining photocurrents from arrayed photomixers with a T-junction power combiner is described using Fig. 2.

The photocurrent I, produced by each UTC-PD, is proportional to the lightwave intensity incident on it and is given as

$$I = I_{dc} + I_{ac}, \tag{2.1}$$

where I_{dc} is the direct current (DC) component, I_{ac} is the alternating current (AC) component that contributes to the THz radiation and is given as

$$I_{ac} = I_{THz} \cos\left(2\pi f_{THz} t + \varphi\right), \tag{2.2}$$

where I_{THz} is the amplitude of the alternating current, f_{THz} and φ are the frequency (e.g. 300 GHz) and the initial phase,

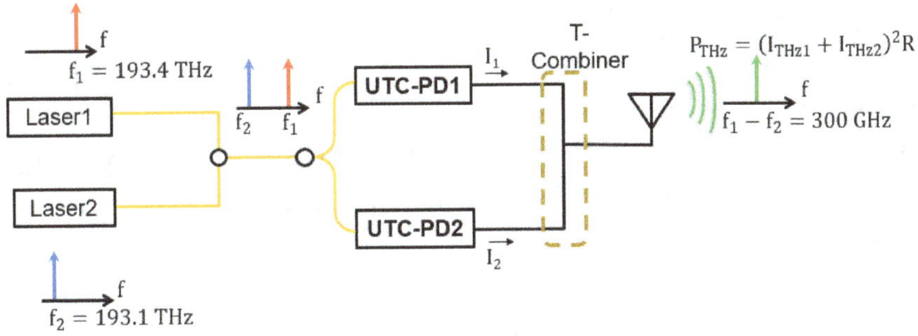

Fig. 2.   Concept of THz wave power enhancement by parallel-connected UTC-PDs.

respectively. If the lightwaves incident on the two UTC-PDs are of the same phase, the T-junction combiner will coherently add the photocurrents from the two UTC-PDs with amplitudes I_{THz1} and I_{THz2}. The radiated power from the antenna will then be

$$P_{THz} = (I_{THz1} + I_{THz2})^2 R, \tag{2.3}$$

where R is the radiation resistance of the antenna.

From the above analysis, it can be deduced that coupling of N lightwaves with identical phases to N combined UTC-PDs leads to an ideal multiplication of the resulting output current by a factor of N and the power of the radiated wave from the antenna is amplified by N^2.

4. Device Design and Fabrication

The T-junction power combiner is a 3-port network and can be implemented in any type of transmission medium such as the MSL and CPW. The T-junction combiner to connect two UTC-PDs in [10] used CPW-based lines. In this work, we employ an MSL-based T-junction combiner which is easy to be cascaded to make a large-scale combiner as well as to be integrated with a patch antenna. As for the device design, at first, the T-junction combiner and the patch antenna on a 32-μm SiC substrate as shown in the inset of Fig. 3 were optimized using a 3D full-wave electromagnetic simulator (Ansys HFSS). The patch antenna was chosen as the THz wave radiator because it radiates the THz wave upward from the substrate. It was based on microstrip patch antenna design equations given in [25–27]. The ports were modeled as lumped ports and were designed to be 0.25-mm apart corresponding to the fiber array pitch. The widths of the 50-Ω line and the quarter wavelength transformer (70.7-Ω) line are 0.031 mm and 0.013 mm, respectively. The optimized patch antenna length, width and the inset feed length are 0.146 mm, 0.213 mm and 0.059 mm, respectively.

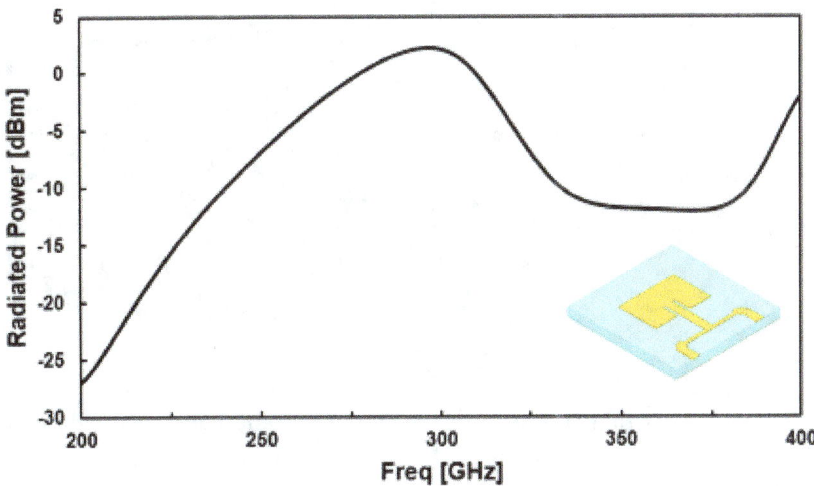

Fig. 3.   Radiated power of the integrated antenna-combiner design simulation results.

The performance of the proposed design was evaluated in terms of radiated power. When 0 dBm (1 mW) was fed to each port of the combiner, as illustrated in Fig. 3, the simulated radiated power at 300 GHz was 2.23 dBm which is 0.77 dB smaller than the total power of 3 dBm (2 mW), with a 3-dB bandwidth of 36.63 GHz. The 0.77-dB loss in the simulated radiated power is potentially due to conductor and dielectric losses which was confirmed by simulation of a straight MSL. It is important to note that in simulation, the signal source was modeled as a power source. This implies that when the number of power sources is doubled, there is ideally a 3-dB (twofold) increase in radiated power. However, if a current source like a UTC-PD is used, doubling the sources would result in a 6-dB (fourfold) increase in radiated power because power is proportional to the square of the current.

In order to avoid RF signal leakage into the DC bias supply, stepped-impedance low-pass filters (LPFs) were utilized in the UTC-PD biasing circuit. The UTC-PD used has a 0.13-μm-thick

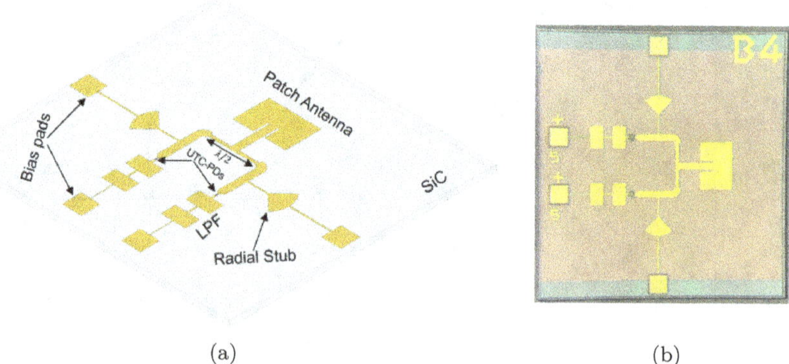

Fig. 4. Device design. (a) Schematic layout of the components. (b) Fabricated device.

InGaAs absorber consisting of p-type and undoped layers and an area of $12.5\,\mu\mathrm{m}^2$ (4-μm diameter). Additionally, it features a 0.23-μm-thick InP depletion layer, similar to the one reported in [9, 23]. The PD was fabricated on a 32-μm-thick SiC substrate using the epi-layer transfer method, as described in [28]. Figure 4 shows the schematic and photo of the fabricated device.

5. Experimental Setup

The experimental setup to characterize the device is shown in Fig. 5. A pair of wavelength-tunable laser diodes, LD1 and LD2, emitted two lightwaves with a power of 10 dBm and frequencies of 193.1 THz and 193.4 THz, respectively. The lightwaves were combined by an optical coupler (OC), modulated at 1 MHz by an intensity modulator (IM) and then amplified by an erbium-doped fiber amplifier (EDFA). The amplified optical signal was split into two optical paths by an optical splitter (OS). The lengths of the two paths were tuned using optical delay lines (ODLs). The two optical signals were then introduced to two UTC-PDs through a microlens array (MLA). At each UTC-PD, an alternating current (AC) with a frequency of 300 GHz was generated. These currents were combined by a T-junction power combiner and fed

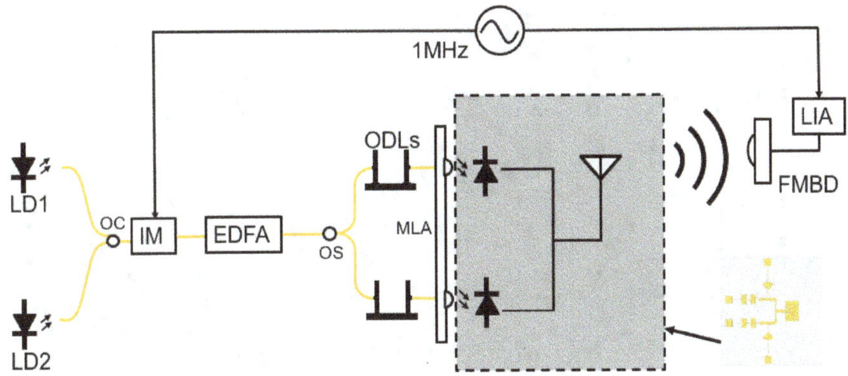

Fig. 5.   Experimental setup: LD: Laser diode, FMBD: Fermi-level managed barrier diode, OC: Optical coupler, IM: Intensity modulator, EDFA: Erbium-doped fiber amplifier, OS: Optical splitter, LIA: Lock-in amplifier, ODLs: Optical delay lines and MLA: Microlens array.

to a rectangular microstrip patch antenna to radiate the 300-GHz wave. The radiated wave was detected using a Fermi-level managed barrier diode (FMBD) [29–31] equipped with a 25-mm-aperture THz lens at 32 mm away from the device. The relative intensity of the 1-MHz component of the detected THz wave was measured using a lock-in amplifier (LIA).

6.   Device Characterization

To assess the feasibility of power enhancement by combining currents at the MSL-based T-junction combiner, the coherency of combined photocurrents, the UTC-PD bias voltage effect on output power, the output linearity as well as the radiation pattern were investigated.

6.1.   *Coherency of combined photocurrent*

In order to achieve coherent signal combining in a T-junction combiner, it is necessary for the phases of the two alternating currents being fed into the combiner to be the same. This means that the

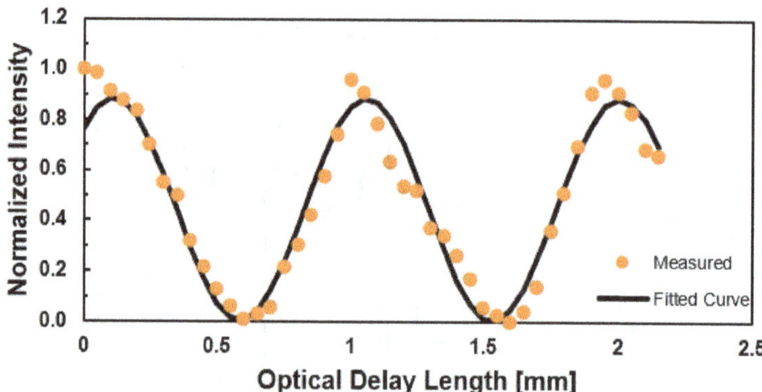

Fig. 6. A 300-GHz wave intensity as a function of delay length in the optical path.

peaks and troughs of the two currents align with each other, resulting in constructive interference and maximum signal combining efficiency. To vary the phases of the two lightwaves traveling in different optical paths, we used a manually operated optical delay line (ODL). By adjusting the length of the optical path, we were able to change the phase of the alternating currents. This allowed us to control the interference pattern that would be observed when the two alternating currents were combined. In Fig. 6, we plotted the power of the 300-GHz wave detected by FMBD. The resulting graph showed a clear interference pattern at a fitted sinusoidal curve. The period of this curve is about 1 mm, which corresponds to the expected interference period between two 300-GHz alternating currents. Also, since the power decreases down to zero, it is confirmed that the T-junction combiner provides an ideal coherent combining.

6.2. *UTC-PD bias voltage effect on output power*

The THz wave output power changes significantly with respect to the bias voltage as shown in Fig. 7. At 300 GHz, the output starts to build up at +0.25-V bias voltage with an optical input power of 20 dBm. As the negative bias voltage increases, there's

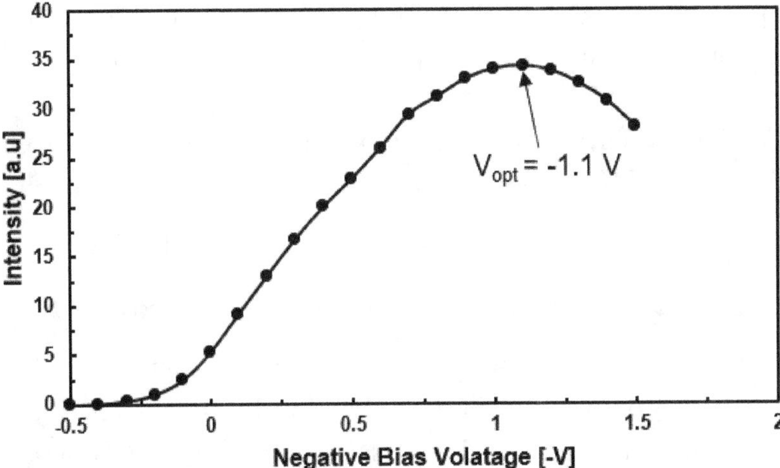

Fig. 7.   Measured THz wave intensity with respect to applied negative bias voltage at 20-dBm optical input power.

a rapid increase in output. The optimal bias voltage (V_{opt}) that produces the highest output is -1.1 V, and beyond this point, the output decreases. It should be noted that at a photocurrent of 7 mA for each UTC-PD, the output is linear and so, the space charge effect is not causing the output degradation beyond V_{opt}. Thus, the output variation is possibly due to the electron velocity overshoot effect, that is sensitive to the electric field intensity [9, 23] whereby a certain range of electric field intensities within the depletion layer can cause electrons to exceed their saturation velocity which affects the response of the UTC-PD, which is determined by the carrier transit time, thereby affecting the output. The optimum field intensity E in a 0.23-μm depletion layer corresponding to a bias voltage V_{bias} of -1.1 V at 300 GHz is calculated to be 80 kV/cm using

$$E = \left(V_b - V_{bias}\right)/W_c, \tag{2.4}$$

where the built-in voltage $V_b = 0.75$ V and the depletion layer width $W_c = 0.23\,\mu$m. This field intensity can lead to an electron velocity between 4×10^7 cm/s and 6×10^7 cm/s. This assertion is in agreement with the experimental and theoretical results reported in [9, 23], where -1.2 V (80 kV/cm), -0.8 V (67 kV/cm) and -0.4 V (50 kV/cm) gave the maximum outputs at 250 GHz, 500 GHz and 1 THz, respectively.

6.3.   *Output linearity with respect to photocurrent*

One of the most important characterization parameters of a UTC-PD-based photomixer for THz application is the output linearity range which alludes to the device saturation output. We assessed our device output linearity by analyzing the relationship between the square of the total photocurrent produced by the UTC-PDs and the output power at a DC bias voltage of -1 V. Before conducting the measurement, the ODL was carefully tuned to ensure optimal output power. To avoid any negative effects or damage to the device, measurements were taken up to an input optical power of 24 dBm, resulting in a photocurrent of 11 mA from each UTC-PD. Figure 8 shows that the results align with the theoretical definition that the output power is proportional to the square of the photocurrent up to around 10 mA from each UTC-PD, assuming the AC current is proportional to the DC current. However, there may be a slight discrepancy due to bias dependency in UTC-PDs, causing variations in the actual bias across the PDs at a fixed external bias voltage as the photocurrent changes. Beyond 20 mA of the total photocurrent from the two UTC-PDs, the output saturates as a result of space charge effect whereby with increasing photocurrent, the current density increases in the depletion layer until it reaches a critical value J_{max} that modulates the field profile therein causing non-linear response and output saturation. The UTC-PD used has an absorption area of $12.5 \, \mu$m^2, hence, the peak photocurrent of 10 mA suggests that the critical current density is 80 kA/cm^2. This is consistent with the

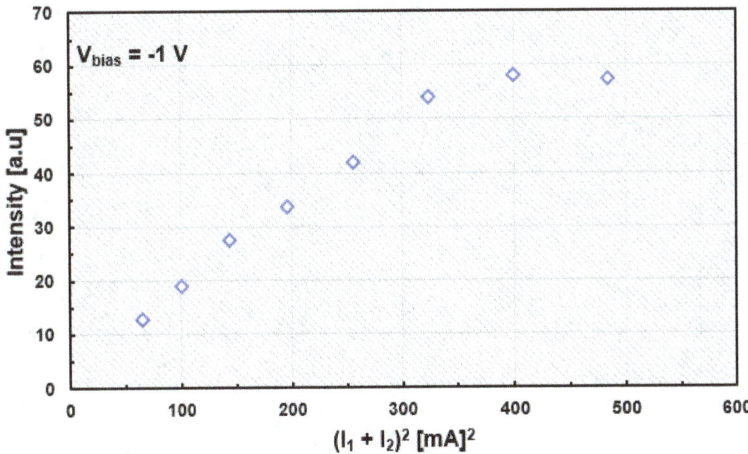

Fig. 8.   THz wave intensity versus the square of the total photocurrent.

theoretical J_{max} estimation given by

$$J_{max} = qn_c v_s, \qquad (2.5)$$

where the electron charge $q = 1.6 \times 10^{-19}$C, carrier density $n_c = 10^{16}$ and v_s can be estimated to be 5×10^7 cm/s.

6.4.   *Radiation pattern*

The power-combining capability of the device was examined at a radiation pattern of 80-deg beam width with a -1 V bias voltage and a photocurrent of 7 mA. As depicted in Fig. 9, when both UTC-PDs were irradiated, the measured power was about 5.8 dB higher relative to a single UTC-PD irradiation over all the angles. These results were consistent with the theoretical explication depicted in Eq. (2.3) that doubling the number of UTC-PDs results in a fourfold increase in the emitted power. The difference of 0.2 dB from the expected power gain of 6 dB is thought to be due to the slight disparity in the photocurrents from the two UTC-PDs. It is important to highlight that this technique of combining THz waves by parallel connection of UTC-PDs to an antenna, allows

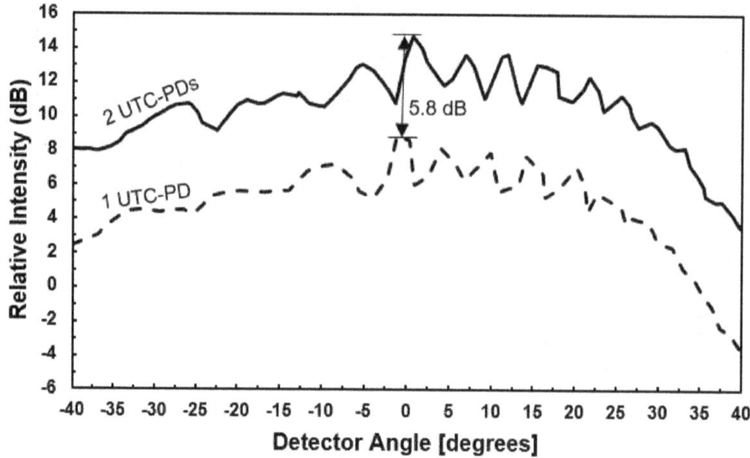

Fig. 9.   Radiation patterns for single and double UTC-PD irradiations.

for isotropic amplification of the THz radiation intensity, making it more efficient in converting optical-to-THz-wave power without necessarily narrowing the beam width compared to using the method of arrayed antennas with UTC-PDs. Although the latter approach is also highly effective in enhancing the peak power of the radiated THz wave, it necessitates an alignment of the beam towards a receiver due to its highly directional radiated pattern.

7.   Conclusion

We presented a comprehensive characterization of a novel fabricated device integrating two parallel-connected UTC-PDs, an MSL-based T-junction combiner and a microstrip patch antenna on an SiC substrate for THz wave power enhancement. It has been shown from the experiment at 300 GHz that the device output is a function of electron velocity overshoot in the depletion layer of the UTC-PD that depends on an optimum bias voltage ($-1.1\,$V) that corresponds to an optimal electric field of $80\,$kV/cm. Moreover, the device saturation depends on space charge effect that is as a result of a critical charge density ($\sim 80\,$kA/cm^2) that modulates the electric field in the depletion region. Furthermore, the results demonstrate a substantial 5.8-dB increase in radiated terahertz power when employing the proposed two UTC-PDs configuration as compared to using a single UTC-PD. It's worth noting that the overall efficacy of the device can be attributed to the lower insertion loss and the simplicity of the T-junction combiner compared to other combiners such as a Wilkinson power combiner that requires a resistor, as well as the efficiency of the microstrip patch antenna. The future goal is to increase the number of UTC-PDs to further enhance the radiated THz power.

Acknowledgments

This work is supported by JST SPRING, Grant Number: JPMJSP2136, and the commissioned research by National Institute of Information and Communications Technology (NICT) JPJ012368C02801 and JPJ012368C00901, the MIC/SCOPE #195010002 and JSPS KAKENHI Grant Numbers: JP21K18730 and JP23K17751. Also, the authors extend their appreciation to Tadao Ishibashi of Wavepackets for the valuable assistance provided throughout the fabrication process.

ORCID

Hussein Ssali https://orcid.org/0009-0006-1944-1860

Yoshiki Kamiura https://orcid.org/0000-0002-6218-2802

Ryo Doi https://orcid.org/0009-0002-6994-7713

Hiroki Agemori https://orcid.org/0009-0005-9685-9016

Ming Che https://orcid.org/0000-0002-8928-4825

Yuya Mikami https://orcid.org/0009-0004-4844-1144

Kazutoshi Kato https://orcid.org/0000-0002-7658-7480

References

1. J. P. Seddon, M. Natrella, X. Lin, C. Graham, C. C. Renaud and A. J. Seeds, *IEEE J. Sel. Top. Quantum Electron.* **28**, 2, 1–12 (2022).
2. H. Sarieddeen, N. Saeed, T. Y. Al-Naffouri and M. S. Alouini, *IEEE Commun. Mag.* **58**, 5, 69–75 (2020).

3. A. Maestrini, I. Mehdi, J. V. Siles, R. Lin, C. Lee, G. Chattopadhyay, J. Pearson and P. Siegel, *Proc. SPIE* **8496**, 84960F (2012).

4. H. Rucker and B. Heinemann, Device architectures for high-speed SiGe HBTs, in *Proc. 2019 IEEE BiCMOS and Compound Semiconductor Integrated Circuits and Technology Symp. (BCICTS)*, Nashville, TN, USA (2019).

5. K. M. K. H. Leong, X. Mei, W. Yoshida, P. Liu, Z. Zhou, M. Lange, L. Lee, J. G. Padilla, A. Zamora, B. S. Gorospe, K. Nguyen and W. R. Deal, *IEEE Microw. Wirel. Compon. Lett.* **25**, 6, 397–399 (2015).

6. H. Hamada, T. Fujimura, I. Abdo, K. Okada, H.-J. Song, H. Sugiyama, H. Matsuzaki and H. Nosaka, 300-GHz. 100-Gb/s InP-HEMT wireless transceiver using a 300-GHz fundamental mixer, in *Proc. 2018 IEEE/MTT-S Int. Microwave Symp. (IMS)*, Philadelphia, PA, USA (2018).

7. I. Mehdi, J. V. Siles, C. Lee and E. Schlecht, *Proc. IEEE* **105**, 6, 990–1007 (2017).

8. H. Ito, F. Nakajima, T. Furuta and T. Ishibashi, *Semicond. Sci. Technol.* **20**, 7, S191 (2005).

9. T. Ishibashi, Y. Muramoto, T. Yoshimatsu and H. Ito, *IEEE J. Sel. Top. Quantum Electron.* **20**, 6, 79–88 (2014).

10. S.-H. Yang, R. Watts, X. Li, N. Wang, V. Cojocaru, J. O'Gorman, L. P. Barry and M. Jarrahi, *Opt. Express* **23**, 24, 31206–31215 (2015).

11. S.-H. Yang and M. Jarrahi, High-power continuous-wave terahertz generation through plasmonic photomixers, in *Proc. 2016 IEEE MTT-S Int. Microwave Symp. (IMS)*, San Fransisco, USA (2016).

12. H. Elayan, O. Amin, B. Shihada, R. M. Shubair and M. S. Alouini, *IEEE Open J. Commun. Soc.* **1**, 1, 1–32 (2019).

13. H.-J. Song, K. Ajito, Y. Muramoto, A. Wakatsuki, T. Nagatsuma and N. Kukutsu, *IEEE Microw. Wirel. Compon. Lett.* **22**, 7, 363–365 (2012).

14. H. Ssali, M. Che and K. Kato, Performance analysis of a Wilkinson power combiner-fed patch antenna for 300-GHz arrayed photomixers, in *Proc. 2022 10th Int. Japan-Africa Conf. Electronics, Communications, and Computations (JAC-ECC)*, Alexandria, Egypt (2022).

15. H. Ssali, M. Che and K. Kato, Coupled line Wilkinson combiner-antenna integrated design for 300-GHz arrayed UTC-PDs, in *Proc. 2023 Opto-Electronics and Communications Conf. (OECC)*, Shanghai, China (2023).

16. Y. Shiratori, T. Hoshi, M. Ida, E. Higurashi and H. Matsuzaki, *IEEE Electron Device Lett.* **39**, 6, 807–810 (2018).

17. M. Watanabe, M. Yanagishita, K. Uesaka, M. Ekawa and H. Shoji, *SEI Tech. Rev.* **83**, 45–49 (2016).

18. T. Sanjoh, N. Sekine, K. Kato, S. Takagi and M. Takenaka, *Jpn. J. Appl. Phys.* **58**, SBBE06 (2019).

19. M. Takenaka and S. Takagi, *Opt. Express* **25**, 29993–30000 (2017).

20. K. Kato, *Photonics* **9**, 1, 9 (2022).

21. M. Che, K. Kondo, H. Kanaya and K. Kato, *J. Lightwave Technol.* **40**, 20, 6657–6665 (2022).
22. T. Ishibashi and H. Ito, Uni-traveling-carrier photodiodes, in *Proc. Technical Digest: Ultrafast Electronics and Optoelectronics*, Lake Tahoe, CA, USA (1997).
23. T. Ishibashi and H. Ito, *J. Appl. Phys.* **127**, 3, 031101-1–031101-10 (2020).
24. T. Ishibashi and H. Ito, *IEEE J. Sel. Top. Quantum Electron.* **28**, 2, 1–6 (2022).
25. C. A. Balanis, *Antenna Theory: Analysis and Design* (John Wiley & Sons, New York, 2005).
26. D. M. Pozar, *Microwave Engineering*, Third Edition (John Wiley & Sons, New York, 2005).
27. T. C. Edwards and M. B. Steer, *Foundations for Microstrip Circuit Design* (Wiley, West Sussex, 2016).
28. H. Ito, N. Shibata, T. Nagatsuma and T. Ishibashi, *Appl. Phys. Express* **15**, 026501 (2022).
29. H. Ito and T. Ishibashi, *Jpn. J. Appl. Phys.* **56**, 1, 014101 (2016).
30. H. Ito and T. Ishibashi, *Electron. Lett.* **51**, 18, 1440–1442 (2015).
31. H. Ito and T. Ishibashi, Novel Fermi-level managed barrier diode for broadband and sensitive terahertz-wave detection, in *Proc. IEEE 40th Int. Conf. Infrared, Millimeter, Terahertz Waves*, Hong Kong, China (2015).

Chapter 3

Temperature Dependence of Hot-Electron Graphene Fet Bolometric Detectors Response to Modulated Terahertz Radiation[#]

Maxim Ryzhii[*,¶], Victor Ryzhii[†,‖], Chao Tang[†], Taiichi Otsuji[†], Vladimir Mitin[‡,], and Michael S. Shur[§,††]**

**Department of Computer Science and Engineering
University of Aizu, Aizu-Wakamatsu,
Fukushima 965-8580, Japan
†Research Institute for Electrical Communication
Tohoku University, Sendai
Miyagi 980-8577, Japan
‡Department of Electrical Engineering
University at Buffalo, SUNY, Buffalo
New York 14260, USA
§Department of Electrical, Computer, and Systems Engineering
Rensselaer Polytechnic Institute, Troy, New York 12180, USA
¶m-ryzhii@u-aizu.ac.jp
‖vryzhii@gmail.com
**vmitin@buffalo.edu
††shurm@rpi.edu*

[#]This chapter appeared previously on the International Journal of High Speed Electronics and Systems. To cite this chapter, please cite the original article as the following: M. Ryzhii, V. Ryzhii, C. Tang, T. Otsuji, V. Mitin and M. S. Shur, *Int. J. High Speed Electron. Syst.*, **33**, 2440024 (2024), doi:10.1142/S012915642440024X.

In this paper, we analyze the modulation characteristics and the ultimate modulation frequency of the terahertz (THz) hot-electron FET bolometers with the graphene channels (GCs), metal gate (MG), and gate barrier layers (BLs) in a wide temperature range. Our results predict that the responsivity of GC-FET bolometers decreases with decreasing operating temperature. This is attributed to a dramatic drop in the thermionic GC-MG current (characterized by the relatively large activation energy) and its modulated component when the temperature lowers. A further decrease in the temperature results in a fairly strong responsivity roll-off. In contrast, the responsivity of the GC-FET detectors with the temperature-adjusted load resistance rises with decreasing temperature because the temperature lowering leads to an increase in the electron energy relaxation time and promotes more effective heating by the impinging THz radiation. In this case, the weakening of the thermionic current is compensated by the commensurate increase in the load resistance.

Keywords: Graphene; terahertz radiation; hot-electron bolometer; field-effect transistor; modulation.

1. Introduction

Recently, we proposed and analyzed the terahertz (THz) hot-electron bolometric detectors based on the field-effect transistors (FETs) with the graphene channel (GC) and the black-AsP (b-AsP) [1, 2] or composite h-BN/b-AsP/h-BN gate barrier layer (BL) [3]. The operation of these GC-FET bolometric detectors is associated with heating the two-dimensional electron gas (2DEG) by the THz radiation, resulting in the reinforcement of the thermionic electron emission from the GC to the metal gate (MG) via the b-AsP gate BL. The selection of the b-AsP as a material for the gate BL is associated with the favorite band alignment at the b-AsP/GC and b-AsP/MG and other useful properties of the b-AsP layer [4–7] (see for details [1–3]). As shown, the resonant excitation of the plasmonic oscillations [8, 9] in the gated GC [10, 11] can enhance the detector responsivity and spectral selectivity. Since the sharpness of the plasmonic resonances increases with decreasing electron collision frequency,

ν, in the GC [10, 11] (i.e., with increasing electron mobility), the GC-FETs with the composite gate BL, in which the GC is encapsulated in h-BN, GC can exhibit marked advantages. Indeed, the room temperature electron mobility in GCs encapsulated in h-BN can be up to an order of magnitude higher than in GCs covered by the b-AsP gate BLs (compare, the data in [12–14], respectively).

The evaluation of the GC-FET detector response to the modulated THz radiation showed that the ultimate modulation frequency can be about a dozen GHz at room temperature [15]. Lowering the working temperature results in a substantial change in the electron momentum and energy relaxation (characterized by the electron collision frequency ν and the electron energy relaxation time τ_ε, in the GC and the electron thermal energy transport along the GC to the side contacts (characterized by the electron thermal conductivity). In line with the Wiedemann–Franz relation for 2DEG in GCs the electron thermal conductivity [16–19] can be presented as $h = v_W^2/2\nu$ [16–19]. A decrease in the specific electron thermal capacitance $c = (2\pi^2 T/3\mu)$ in the degenerate 2DEG in GCs with decreasing temperature (see, for example [16, 17]) affects the rate of electron heating by the impinging THz radiation. The trade-off between the electron cooling in the GC and at its side contacts affects both the responsivity and modulation characteristics of the GC-FET detectors.

In this paper, we study how the lowering of operation temperature can affect the GC-FET detector modulation characteristics, in particular, the ultimate modulation frequency.

2. GC-FET Bolometric Detector Model

Figures 1 and 2 show the structures and the band diagrams of the GC-FET detector with the uniform [1, 2] and composite [3, 15] gate BLs at the bias gate voltage V_G. The electron heat transport equation (for details, see [2, 3]) describes the variation of the

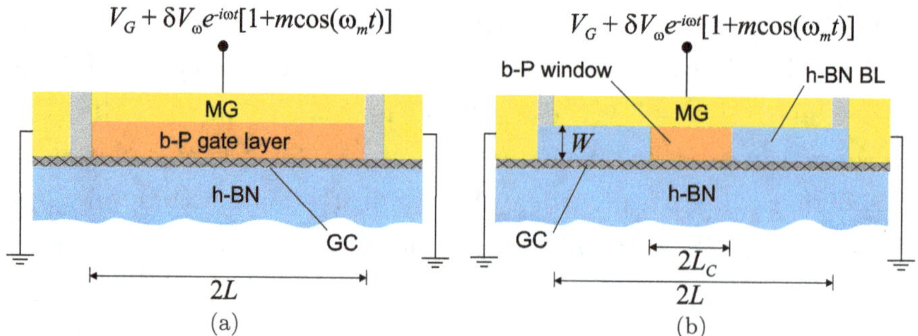

Fig. 1. (a) Cross-section of the GC-FET detector structure with (a) uniform b-P and (b) composite h-BN/black-AsP/h-BN gate BLs.

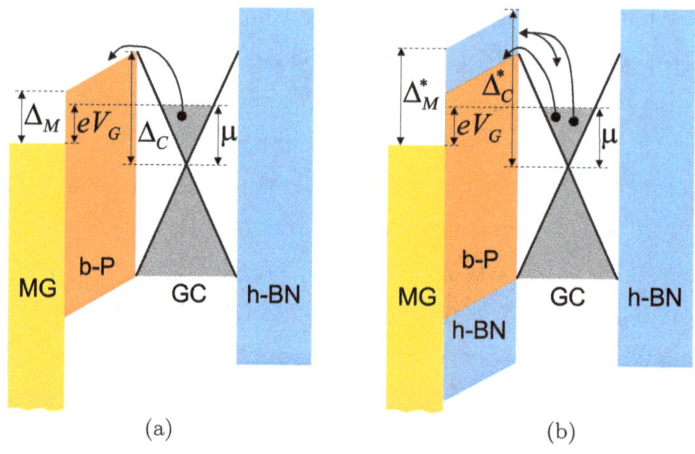

Fig. 2. Band diagram (a) of the GC-FET with uniform b-P gate BL, and (b) of the GC-FET with composite gate with its b-P central section (emission window, $|x| < L_C$) and h-BN wide gap sections ($\Delta_C^* > \Delta_C$) adjacent to the side contacts, $L - L_C < |x| < L$.

2DEG effective electron temperature determined by its heating by the THz radiation received by an antenna and the cooling due to the electron energy relaxation in the GC and at the side contacts. The excitation of plasmonic oscillations in the gated GC is considered in the framework of the 2DEG transport hydrodynamic equations coupled with the Poisson equation. In principle, both detectors are described by the same formulas (but with different values of ν). Due to this, we focus on the GC-FET detectors with the composite gate BL.

In the GC-FET structures with the 2DEG Fermi energy $\mu = \Delta_C - \Delta_M$, where Δ_C and Δ_M are heights of the barriers for the electrons at the b-AsP/GC and b-AsP/MG interfaces [see Fig. 2(a)], the flat-band conditions are realized at $V_G = 0$. In such a situation, the bias voltage, which provides the electron thermionic current mainly from the GC into the MG can be relatively small $(eV \gtrsim T)$, and this voltage induces a relatively small extra electron density in the GC (compared with the donor density). Hence, the electron Fermi energy μ is virtually independent of V_G. This is primarily assumed in the following.

For the definiteness, we focus on the GC-FET detectors based on the structures with the n-BN/b-P/h-BN gate BLs with Al MG assuming that $\Delta_C = 225\,\mathrm{meV}$ and $\Delta_M = 85\,\mathrm{meV}$ (see [1] and the references therein). The 2DEG Fermi energy, corresponding to the flat-band condition at $V_G = 0$, equals to $\mu = 140\,\mathrm{meV}$. Accounting for the relation between the Fermi energy μ and the electron density Σ in the degenerate 2DEG $\Sigma \simeq \mu^2/\pi\hbar^2 v_W^2$, we find that the above Fermi energy corresponds to the 2DEG density $\Sigma \simeq 1.6 \times 10^{12}\,\mathrm{cm}^{-2}$. The chemical doping of the GC or the h-BN substrate might be necessary to achieve such a density at relatively moderate gate voltages.

We assume that the impinging THz radiation with the carrier frequency ω is modulated by the signal with the intensity $I_\omega^{\omega_m} \cos(\omega_m t)$ and the modulation frequency ω_m.

The detector voltage responsivity, $U_\omega^{\omega_m} = \langle J_\omega^{\omega_m} \rangle \rho_L / I_\omega^{\omega_m}$ (in the V/W units), where ρ_L is the load resistance and $\langle J_\omega^{\omega_m} \rangle$ is the amplitude of terminal current averaged over the period $2\pi/\omega$), can be presented as [3, 15]

$$U_\omega^{\omega_m} \propto \frac{\rho_L L_C H}{\tau_\perp} \frac{\Delta_M}{T} \exp\left(-\frac{\Delta_M}{T}\right) \left(\frac{v_W^2 \tau_\varepsilon}{L^2 \nu}\right) \frac{r_\omega |\Pi_\omega^{\omega_m}|}{\sqrt{1 + (\omega_m/\overline{\omega}_m)^2}},$$

$$(3.1)$$

with $2L$, $2L_C$, H, and T being the GC length, the length of the b-P window, the width of the GC-FET structure (i.e., GC), and the lattice temperature, respectively, and $\tau_\perp$ is the electron try-to-escape time from the GC into the MG. Here

$$r_\omega = \frac{1}{[\sin^2(\pi\omega/2\Omega) + (4\Omega/\pi\nu)2\cos^2(\pi\omega/2\Omega)]}, \qquad (3.2)$$

is the factor associated with the plasmonic response of the gated 2DEG in the GC to the THz radiation received by an antenna and $\Omega = (\pi e/\hbar L)\sqrt{\mu W/\kappa}$ with W and κ being the thickness of the gate BL and its dielectric constant. The substrate has the same dielectric constant. The factor

$$\Pi_\omega^{\omega_m} \simeq 1 - \frac{1}{1 + (\pi\omega/a_m\Omega)^2} - \left[1 - \frac{\cos(\pi\omega/\Omega)}{1 + (\pi\omega/a_m\Omega)^2}\right]\frac{1}{\cosh(a_m)}, \qquad (3.3)$$

is associated with the effect of the side contacts on the electron effective temperature spatial distributions along the GL, where the parameters $a_m = a\sqrt{1 - i\omega_m/\overline{\omega}_m}$, $a = L\sqrt{2\nu/v_W^2\tau_\varepsilon}$, and $\overline{\omega}_m = 1/c\tau_\varepsilon$.

For the load resistance ρ_L being equal to the GC-MG resistance, ρ_0, at T equal to the room temperature $T_0 = 25\,\mathrm{meV} \simeq 300\,\mathrm{K}$ and $eV_G = T_0$, we obtain

$$\rho_L \simeq \frac{\tau_\perp T_0}{2L_C H e^2 \Sigma}\exp\left(\frac{\Delta_M}{T_0}\right)$$

$$\simeq \frac{\pi}{2HL_C}\exp\left(\frac{\Delta_M}{T_0}\right)\left(\frac{\hbar v_W}{\mu}\right)^2\left(\frac{\tau_\perp T_0}{e^2}\right). \qquad (3.4)$$

Then, using Eqs. (3.1)–(3.4), for the responsivity at the carrier frequency corresponding to the plasmonic resonance $\omega = \Omega$, we arrive at

$$U_\Omega^{\omega_m} = U_0 \exp\left[-\frac{\Delta_M}{T_0}\left(\frac{T_0}{T}-1\right)\right]$$

$$\times \frac{\left|1 - \frac{1}{1+(\pi/a_m)^2} - \left[1 + \frac{1}{1+(\pi/a_m)^2}\right]\frac{1}{\cosh(a_m)}\right|}{a^2\sqrt{1+(\omega_m/\overline{\omega}_m)^2}}, \qquad (3.5)$$

with U_0 being the characteristic voltage responsivity at $T = T_0$ given by

$$U_0 = \frac{64}{137g}\left(\frac{\hbar}{eT_0}\right)\left(\frac{\Delta_M}{\mu}\right), \qquad (3.6)$$

where $g \simeq 1.6$ is the antenna gain for a dipole antenna. For the device parameters listed above and $T_0 = 25\,\mathrm{meV}$, Eq. (3.5) yields $U_0 \simeq 2.76 \times 10^4\,\mathrm{V/W}$.

At elevated modulation frequencies $2\omega_m \gg \overline{\omega}_m/a^2$, Eq. (3.5) yields

$$U_\Omega^{\omega_m} \simeq U_0 \exp\left[-\frac{\Delta_M}{T_0}\left(\frac{T_0}{T}-1\right)\right]\frac{\pi^2}{a^4}\left(\frac{\overline{\omega}_m}{\omega_m}\right)^2 \propto \left(\frac{\overline{\omega}_m}{\omega_m}\right)^2.$$

$$(3.7)$$

If the load resistance is adjusted to the detector resistance at operating temperature T, its value at $eV_G = T_0$ is given by

$$\tilde{\rho}_L \simeq \rho_L \exp\left[\frac{\Delta_M}{T_0}\left(\frac{T_0}{T}-1\right)\right]. \qquad (3.8)$$

In this regime, for the voltage responsivity $\tilde{U}_\Omega^{\omega_m}$ we obtain [instead of Eq. (3.5)]

$$\tilde{U}_\Omega^{\omega_m} = U_0\frac{\left|1 - \frac{1}{1+(\pi/a_m)^2} - \left[1 + \frac{1}{1+(\pi/a_m)^2}\right]\frac{1}{\cosh(a_m)}\right|}{a^2\sqrt{1+(\omega_m/\overline{\omega}_m)^2}}. \qquad (3.9)$$

Comparing Eqs. (3.5) and (3.9), one can see that due to $\tilde{\rho}_L > \rho_L$ at $T < T_0$, at lowered temperatures $\tilde{U}_\Omega^{\omega_m} > U_\Omega^{\omega_m}$ (see below).

Equations (3.5) and (3.9) lead to the following formula for the GC-FET detector modulation responsivity efficiency at $\omega \simeq \Omega$ defined as $M_\Omega^{\omega_m} = U_\Omega^{\omega_m}/U_\Omega^0 = \tilde{U}_\Omega^{\omega_m}/\tilde{U}_\Omega^0$:

$$M_\Omega^{\omega_m} = \frac{1}{\sqrt{1+(\omega_m/\overline{\omega}_m)^2}} \left| \frac{1 - \frac{1}{1+(\pi/a_m)^2} - \left[1+\frac{1}{1+(\pi/a_m)^2}\right]\frac{1}{\cosh(a_m)}}{1 - \frac{1}{1+(\pi/a)^2} - \left[1+\frac{1}{1+(\pi/a)^2}\right]\frac{1}{\cosh(a)}} \right|. \tag{3.10}$$

3. Temperature Dependences of the Electron Scattering and Energy Relaxation Parameters

As follows from the above section, the performance of the GC-FET bolometric detectors, particularly its dependence on the operating temperature, is determined by the temperature dependences of the electron collision frequency ν and the electron energy relaxation time τ_ε in the GC.

At room temperature and not particularly low temperatures, the electron momentum relaxation time is determined primarily by the acoustic phonon scattering so that the collision frequency given by [20, 21]

$$\nu = \nu_0 \left(\frac{T}{T_0}\right), \tag{3.11}$$

where ν_0 is the frequency of the electron collisions with acoustic phonons at room temperature T_0. In the same temperature range, the electron energy relaxation is associated with the optical phonon scattering [22–24]. The pertinent energy relaxation time τ_ε^{op} expressed via its value at room temperature can be presented as [25, 26]

$$\tau_\varepsilon^{\mathrm{op}} = \tau^{\mathrm{op}} \left(\frac{T}{T_0}\right)^2 \exp\left[\frac{\hbar\omega_0}{T_0}\left(\frac{T_0}{T}-1\right)\right]. \tag{3.12}$$

Here $\hbar\omega_0 \simeq 200\,\mathrm{meV}$ is the optical phonon energy. At low temperatures, the optical phonon scattering practically vanishes, being replaced by the acoustic phonon scattering. In this case, the energy relaxation time $\tau_\varepsilon^{\mathrm{ac}}$, associated with the acoustic scattering, can be presented as (for example [20, 21])

$$\tau_\varepsilon^{\mathrm{ac}} = \tau^{\mathrm{ac}} \left(\frac{T_0}{T} \right), \tag{3.13}$$

where $\tau^{\mathrm{ac}} \gg 1/\nu_0$ corresponds to room temperature.

To consider a sufficiently wide temperature range in line with Eqs. (3.12) and (3.13), we use the following interpolation formula:

$$\frac{1}{\tau_\varepsilon} = \frac{1}{\tau^{\mathrm{op}}} \left(\frac{T_0}{T} \right)^2 \exp\left[\frac{\hbar\omega_0}{T_0} \left(1 - \frac{T_0}{T} \right) \right] + \frac{1}{\tau^{\mathrm{ac}}} \left(\frac{T}{T_0} \right). \tag{3.14}$$

Consequently, for the parameters a and $\overline{\omega}_m$, we have

$$a = a_0 \sqrt{ \frac{\tau_0}{\tau^{\mathrm{op}}} \left(\frac{T_0}{T} \right) \exp\left[\frac{\hbar\omega_0}{T_0} \left(1 - \frac{T_0}{T} \right) \right] + \frac{\tau_0}{\tau^{\mathrm{ac}}} \left(\frac{T}{T_0} \right)^2 },$$

$$\tag{3.15}$$

$$\overline{\omega}_m = \overline{\omega}_m^0 \left\{ \frac{\tau_0}{\tau^{\mathrm{op}}} \left(\frac{T_0}{T} \right)^3 \exp\left[\frac{\hbar\omega_0}{T_0} \left(1 - \frac{T_0}{T} \right) \right] + \frac{\tau_0}{\tau^{\mathrm{ac}}} \right\}, \tag{3.16}$$

with $a_0 = L\sqrt{2\nu_0/v_W^2\tau_0}$ and $\overline{\omega}_m^0 = 1/c_0\tau_0$, where $\tau_0 = \tau^{\mathrm{op}}\tau^{\mathrm{ac}}/(\tau^{\mathrm{op}} + \tau^{\mathrm{ac}}) \simeq \tau^{\mathrm{op}}$ and $c_0 = (2\pi^2 T_0/3\mu)$.

4. Effect of Temperature on Responsivity and Modulation Efficiency

Accounting for the temperature dependences of ν, τ_ε, a, and $\overline{\omega}$ given by Eqs. (3.11)–(3.16) and using Eq. (3.5), (3.9) and (3.10), we calculated the GC-FET detector responsivity and modulation efficiency at $\omega = \Omega$. Apart from the values of the parameters Δ_M and μ assumed above, we set $\nu_0 = 3\,\mathrm{ps}^{-1}$, $\tau^{\mathrm{op}} = 10\,\mathrm{ps}$, $\tau^{\mathrm{ac}} = 1\,\mathrm{ns}$, and $\hbar\omega_0/T_0 = 8$, so that $a_0 \simeq L[\mu m]\sqrt{3/5}$ and $\overline{\omega}_m^0/2\pi = 13.5\,\mathrm{GHz}$.

Figures 3 and 4 show the GC-FET responsivities $U_\Omega^{\omega_m}$ and $\tilde{U}_\Omega^{\omega_m}$ as function of the modulation frequency $\omega_m/2\pi$ for different temperatures T and different values of the GC lengths $2L$.

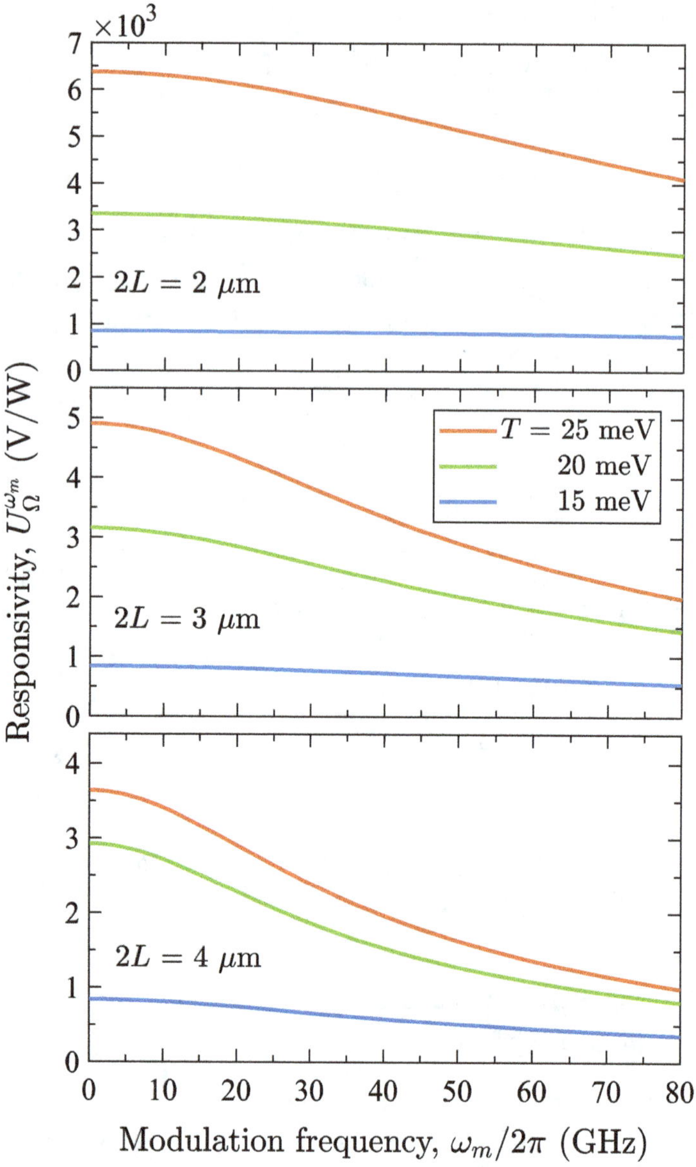

Fig. 3. Responsivity $U_\Omega^{\omega_m}$ versus the modulation frequency $\omega_m/2\pi$ for the GC-FETs with different GC length $2L$ at different temperatures T.

As seen from Fig. 3, the responsivity $U_\Omega^{\omega_m}$ decreases with decreasing operating temperature T. This is attributed to a dramatic drop in the thermionic GC-MG current (characterized by the relatively large activation energy) and its modulated component

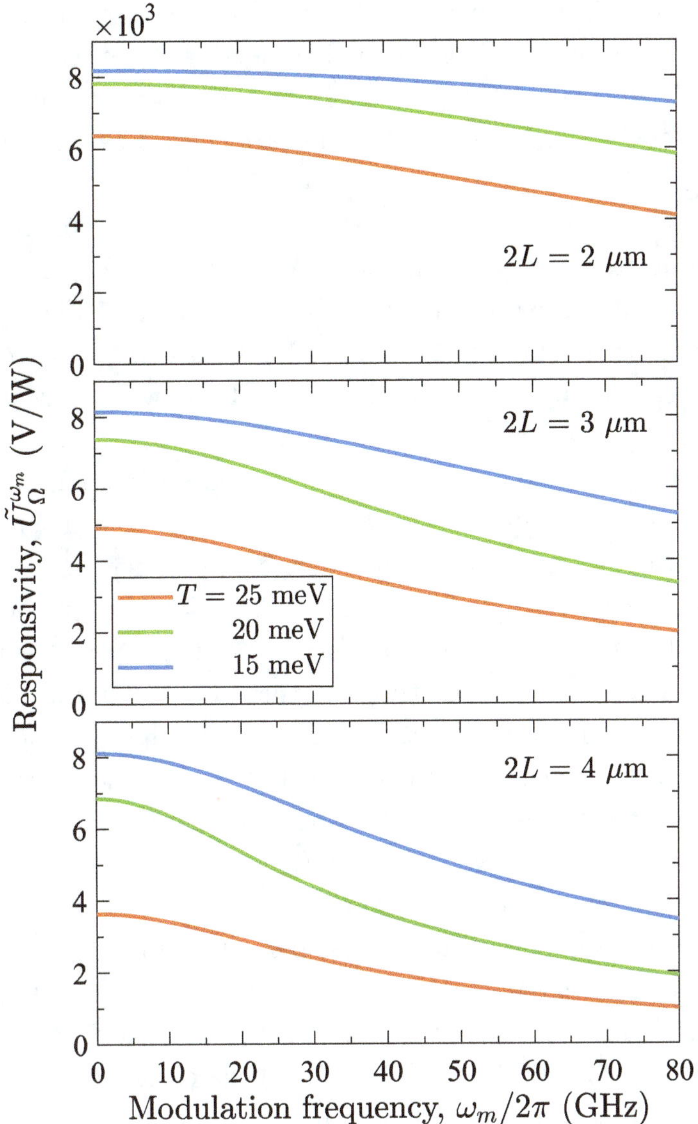

Fig. 4.   The same as in Fig. 3, but for $\tilde{U}_\Omega^{\omega_m}$.

when the temperature lowers. Further decrease in the temperature results in a fairly strong $U_\Omega^{\omega_m}$ roll-off.

In contrast, the responsivity of the GC-FET detectors with the temperature-adjusted load resistance $\tilde{U}_\Omega^{\omega_m}$ rises with decreasing temperature, as shown in Fig. 4. This is because the temperature lowering leads to an increase in the electron energy relaxation time. This promotes more effective heating by the impinging THz radiation, while the weakening of the thermionic current is assumed to be compensated by the proper increase in the load resistance.

Both $U_\Omega^{\omega_m}$ and $\tilde{U}_\Omega^{\omega_m}$, being rather large in a wide range of the modulation frequency, are smaller in the GC-FET detectors with longer GC (compare the dependences on the panels in Figs. 3 and 4 corresponding to different $2L$. This is explained by a weaker THz electric field in longer GCs (at the same THz power received by an antenna).

Figure 5 shows the $M_\Omega^{\omega_m}$ versus modulation frequency $\omega_m/2\pi$ dependences calculated for the different GC length $2L$. This characteristic is more universal because it is independent of the load resistance. For the calculations of the $M_\Omega^{\omega_m}$ frequency characteristics, the same parameters as for Figs. 3 and 4 were assumed.

The most important features of these characteristics seen in Fig. 5 are as follows. First, the $M_\Omega^{\omega_m}$ roll-off with increasing modulation frequency ω_m is remarkably slow. This corresponds to relatively high values of the ultimate modulation frequency defined by the equation $M_\Omega^{\omega_m^{\max}} = 1/\sqrt{2}$. For the GC-FET detectors with the chosen device parameters $\omega_m^{\max}/2\pi$ is in the range of dozens GHz. Second, lowering of the operating temperature leads to slowing down of the $M_\Omega^{\omega_m}$ dependence on the modulation frequency ω_m and, hence, to larger values of the ultimate modulation frequency $\omega_m^{\max}/2\pi$. Third, the GC-FET detectors with a longer GC are characterized by smaller values of $\omega_m^{\max}/2\pi$.

For $\mu = 140\,\mathrm{meV}$, $W = 10^{-5}\,\mathrm{cm}$, and $\kappa \simeq 4$, the plots in Figs. 3–5 correspond to the resonant plasmonic frequencies $\Omega/2\pi \simeq 1.8, 1, 2$ and $0.9\,\mathrm{THz}$, respectively. The thickness of the b-P gate BL in the emission window W_C can be different from the BL thickness W in the side section.

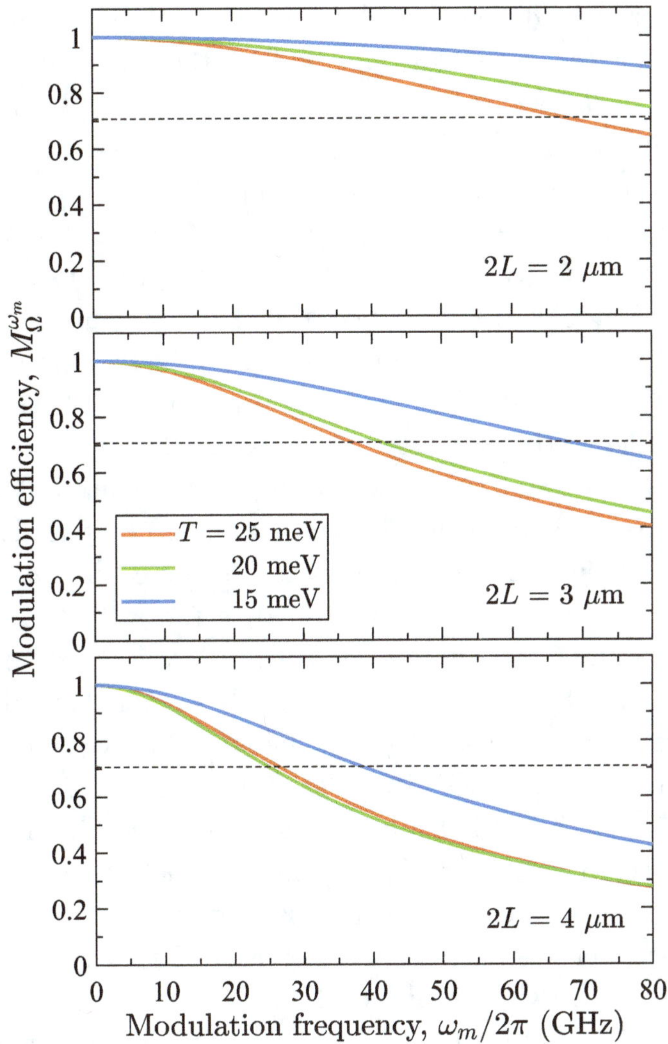

Fig. 5. Frequency dependence of modulation efficiency $M_\Omega^{\omega_m}$ for the GC-FETs with different GC length $2L$ at different temperatures.

5. Comments

(i) Above we set $\nu \propto T$. This corresponds to the acoustic phonon and point defect electron scattering. The electron scattering rate on the ionized impurities yields a weakly decreasing temperature dependence [27]. However, these temperature dependences are weak compared with the exponential dependence of the energy relaxation time due to the optical phonon scattering.

(ii) For $\Delta_M = 85\,\mathrm{meV}$, $\Sigma = 1.6 \times 10^{12}\,\mathrm{cm}^{-2}$ ($\mu = 140\,\mathrm{meV}$), $2L_C = (0.25 - 0.50)\,\mu\mathrm{m}$, $H = 10\mu$, and $\tau_\perp = 10\,\mathrm{ps}$, the GC-MG resistance at $eV_G \sim T_0 = 25\,\mathrm{meV}$ is estimated as $\rho \simeq (65 - 130)$ Ohm. The required load resistance can vary depending on the GC-FET structure width H. This resistance can be decreased using the periodic multiple GC-FET structures with the FETs connected in in parallel. Since the current responsivity and the dark current of the multiple GC-FET detectors are proportional to the number of the periods N, the dark-current-limited detectivity of such detectors $D^* \propto \sqrt{N}$.

(iii) The comparison of the GC-FET detectors with the uniform BL [see Fig. 1(a)] and the composite BL [see Fig. 1(b)] shows that these devices exhibit qualitatively similar characteristics. However, due to a relative smallness of the electron collision frequency ν in the GC-FET detectors with the composite BL (ν in the h-BN BL main part is smaller than it in the b-P BL), the plasmonic resonances in these devices can be more pronounced leading to higher responsivity.

(iv) The model used above assumes that the carrier radiation frequency ω (which is in the THz range) substantially exceeds the modulation frequency ω_m. This assumption was used for the correct averaging of the terminal current over the period $2\pi/\omega$. Hence, the obtained expressions for the GC-FET detector characteristics are valid for $\omega_m/2\pi \lesssim 100\,\mathrm{GHz}$. The latter was accounted for in the plots of Figs. 3–5.

6. Conclusion

We evaluated the modulation characteristics of the hot-electron GC-FET bolometric detectors with graphene channel and the composite h-BN/b-P barrier in a wide temperature range for modulation frequencies of dozens GHz. Our results allow choosing the temperature-dependent matching load resistances for the optimum bolometer performance.

Acknowledgments

The work was supported: at UA and TU - by the Japan Society for Promotion of Science (KAKENHI Nos. 21H04546 and 20K20349), Japan, and by the RIEC Nation-Wide Collaborative Research Project No. R04/A10, Japan, and at RPI - by AFOSR (contract number FA9550-19-1-0355).

ORCID

Maxim Ryzhii https://orcid.org/0000-0001-5580-5971

Victor Ryzhii https://orcid.org/0000-0002-5000-3109

Chao Tang https://orcid.org/0000-0002-6788-9575

Taiichi Otsuji https://orcid.org/0000-0002-0887-0479

Vladimir Mitin https://orcid.org/0000-0002-5579-5803

Michael S. Shur https://orcid.org/0000-0003-0976-6232

References

1. V. Ryzhii, C. Tang, T. Otsuji, M. Ryzhii, V. Mitin and M. S. Shur, "Resonant plasmonic detection of terahertz radiation in field-effect transistors with the graphene channel and the black-As_xP_{1-x} gate layer," *Sci. Rep.* **13** (2023) 9665.
2. V. Ryzhii, C. Tang, T. Otsuji, M. Ryzhii, V. Mitin and M. S. Shur, "Effect of electron thermal conductivity on resonant plasmonic detection in the metal/black-AsP/graphene FET terahertz hot-electron bolometers," *Phys. Rev. Appl.* **19** (2023) 064033.
3. M. Ryzhii, V. Ryzhii, M. S. Shur, V. Mitin, C. Tang and T. Otsuji, "Terahertz bolometric detectors based on graphene field-effect transistors with the composite h-BN/black-P/h-BN gate layers using plasmonic resonances," *J. Appl. Phys.* **134** (2023) 084501.
4. Y. Cai, G. Zhang and Y.-W. Zhang, "Layer-dependent band alignment and work function of few-layer phosphorene," *Sci. Rep.* **4** (2015) 6677.
5. Xi Ling, H. Wang, S. Huang, F. Xia and M. S. Dresselhaus, "The renaissance of black phosphorus," *Proc. Natl. Acad. Sci.* **122** (2015) 4523–4530.

6. B. Liu, M. Köpf, A. N. Abbas, X. Wang, Q. Guo, Y. Jia, F. Xia, R. Weihrich, F. Bachhuber, F. Pielnhofer, H. Wang, R. Dhall, S. B. Cronin, M. Ge, X. Fang, T. Nilges and C. Zhou, "Black Arsenic–Phosphorus: Layered anisotropic infrared semiconductors with highly tunable compositions and propertie," *Adv. Mater.* **27** (2015) 4423–4429.

7. M. Uda, A. Nakamura, T. Yamamoto and Y. Fujirnoto, "Work function of polycrystalline Ag, Au and Al," *J. Electron. Spectros. Relat. Phenomena* **88–91** (1998) 643–648.

8. M. S. Shur and V. Ryzhii, "Plasma wave electronics," *Int. J. High Speed Electron. Syst.* **13** (2003) 575–600.

9. V. Ryzhii, I. Khmyrova, M. Ryzhii, A. Satou, T. Otsuji, V. Mitin and M. S. Shur, "Plasma waves in two-dimensional electron systems and their applications," *Int. J. High Speed Electron. Syst.* **17** (2007) 521–538.

10. V. Ryzhii, A. Satou and T. Otsuji, "Plasma waves in two-dimensional electron-hole system in gated graphene heterostructures," *J. Appl. Phys.* **101** (2007) 024509.

11. V. Ryzhii, T. Otsuji and M. S. Shur, "Graphene based plasma-wave devices for terahertz applications," *Appl. Phys. Lett.* **116** (2019) 140501.

12. A. S. Mayorov, R. V. Gorbachev, S. V. Morozov, L. Britnell, R. Jalil, L. A. Ponomarenko, P. Blake, K. S. Novoselov, K. Watanabe, T. Taniguchi and A. K. Geim, "Micrometer-scale ballistic transport in encapsulated graphene at room temperature," *Nano Lett.* **11** (2011) 2396–2399.

13. M. Yankowitz, Q. Ma, P. Jarillo-Herrero and B. J. Le Roy, "Van der Waals heterostructures combining graphene and hexagonal boron nitride," *Natl. Rev. Phys.* **1** (2019) 112–125.

14. Y. Liu, I. Yudhistira, M. Yang, E. Laksono, Y. Z. Luo, J. Chen, J. Lu, Y. P. Feng, S. Adam and K. P. Loh, "Mediated colossal magnetoresistance in graphene/black phosphorus heterostructures," *Nano Lett.* **18** (2018) 3377.

15. V. Ryzhii, C. Tang, T. Otsuji, M. Ryzhii, V. Mitin and M. S. Shur, "Dynamic characteristics of terahertz hot-electron graphene FET bolometers: Effect of electron cooling in channel and at side contacts," preprint (2024), arXiv: 2403.06373.

16. Q. Li, X. C. Xie and S. Das Sarma, "Calculated heat capacity and magnetization of two-dimensional electron systems," *Phys. Rev. B* **40** (1989) 1381(R).

17. M. Massicotte, G. Soavi, A, Principi and K.-J. Tielrooij, "Hot carriers in graphene – fundamentals and applications," *Nanoscale* **13** (2021) 8376–8411.

18. Z. Tong, A. Pecchia, C. Yam, T. Dumitric and T. Frauenheim, "Ultrahigh electron thermal conductivity in T-Graphene, Biphenylene, and Net-Graphene," *Adv. Energy Mater.* **12** (2022) 2200657.

19. T. Y. Kim, C.-H. Park and N. Marzari, "The electronic thermal conductivity of graphene," *Nano Lett.* **16** (2016) 2439–2443.

20. F. T. Vasko and V. Ryzhii, "Voltage and temperature dependences of conductivity in gated graphene heterostructures," *Phys. Rev. B* **76** (2007) 233404.

21. E. H. Hwang and S. Das Sarma, "Acoustic phonon scattering limited carrier mobility in two-dimensional extrinsic graphene," *Phys. Rev. B* **77** (2008) 115449.
22. J. H. Strait, H. Wang, S. Shivaraman, V. Shields, M. Spencer and F. Rana, "Very slow cooling dynamics of photoexcited carriers in graphene observed by optical-pump terahertz-probe spectroscopy," *Nano Lett.* **11** (2011) 4902.
23. V. Ryzhii, M. Ryzhii, V. Mitin, A. Satou and T. Otsuji, " Effect of heating and cooling of photogenerated electron-hole plasma in optically pumped graphene on population inversion," *Japan J. Appl. Phys.* **50** (2011) 094001.
24. K. Tamura, C. Tang, D. Ogiura, K. Suwa, H. Fukidome, Y. Takida, H. Minamide, T. Suemitsu, T. Otsuji and A. Satou, "Fast and sensitive terahertz detection in a current-driven epitaxial-graphene asymmetric dual-grating-gate FET structure," *APL Photonics* **7** (2022) 126101.
25. V. Ryzhii, T. Otsuji, M. Ryzhii, M. Ryzhii, N. Ryabova, S. O. Yurchenko, V. Mitin and M. S. Shur, "Graphene terahertz uncooled bolometers," *J. Phys. D: Appl. Phys.* **46** (2013) 065102.
26. V. Ryzhii, A. Satou, T. Otsuji, M. Ryzhii, V. Mitin and M. S. Shur, " Graphene vertical hot-electron terahertz detectors," *J. Appl. Phys.* **116** (2014) 114504.
27. S. K. Jaćimovski, J. S. Lamovec, J. P. Šetrajčić and D. I. Ilić, "Temperature dependence of the relaxation time in scattering of elementary excitations in graphene," *Math. Sci. Eng. B* **264** (2021) 114933.

Chapter 4

Sensing Using Terahertz Radiation[#]

Michael Shur

*Department of Electrical, Computer and Systems Engineering,
Rensselaer Polytechnic Institute, 110 8th Street, Troy,
NY 12180, USA*

shurm@rpi.edu

Terahertz (THz) sensing technology enables 6G communication, detection of biological and chemical hazardous agents, cancer detection, monitoring of industrial processes and products, and detection of mines and explosives. THz sensors support security in buildings, airports, and other public spaces. They found important applications in radioastronomy and space research and, more recently, in Artificial Intelligence-driven THz sensing of MMICs and VLSI. Exploding demand for data transfers will require using the 300 GHz band after 2028 or even before and will make the deployment of THz sensing electronics inevitable. This paper discusses the new physics of THz sensing and THz sensing devices. It also reviews the THz sensing market, and key THz sensor companies.

Keywords: Sensing; terahertz; TeraFETs; plasmonics; 6G communication.

[#]This chapter appeared previously on the International Journal of High Speed Electronics and Systems. To cite this chapter, please cite the original article as the following: M. Shur, *Int. J. High Speed Electron. Syst.*, **33**, 2440022 (2024), doi:10.1142/S0129156424400226.

1. Introduction: THz Gap and Counter-Intuitive Physics of THz Sensing

The use of electromagnetic radiation and artificial energy sources is a prerequisite for human civilization. Chinese scientists are saying an early human ancestor, Peking Man, set up fireplaces and cooked food about 600,000 years ago — the earliest evidence of fire use by a human species. Figure 1 shows the timeline of the evolution of artificial sources of electromagnetic radiation from that time to the terahertz (THz) technology range.

The Argand lamp inventor, François-Pierre-Amédée Argand was a student of Antoine Lavoisier who was the first researcher of lighting studying the effects of the combustion of oxygen on lighting [1]. Tragically, he was guillotined during the French Revolution. The myth is that staying a researcher till the bitter end Antoine Lavoisier asked a friend to count how many times he would blink after his execution (the story goes that he blinked 14 times).

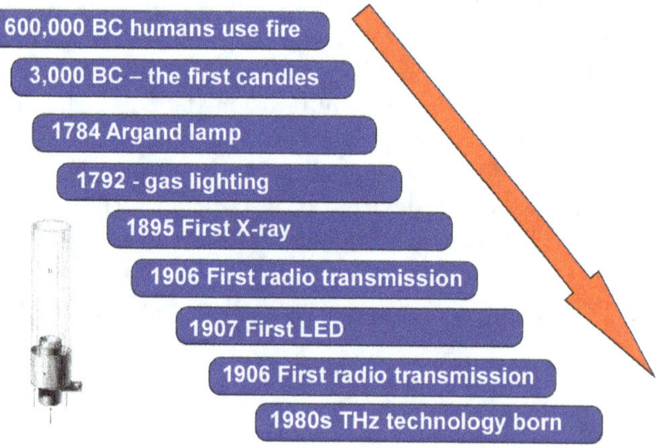

Fig. 1.   The timeline of the evolution of artificial sources of electromagnetic radiation from using fire to terahertz (THz) technology. The inset shows the Agrand lamp (source Library of Congress Photo Library) — the first lamp based on research.

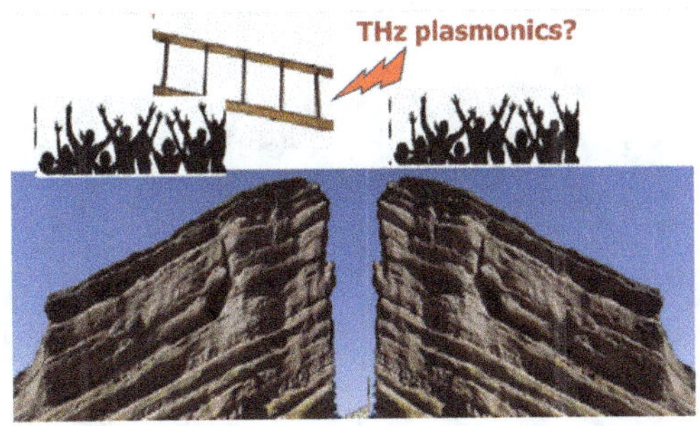

Fig. 2.   THz Gap to be bridged by THz plasmonics.

Figure 2 illustrates the famous THz gap and the hope that it will be bridged using THz plasmonic which we discuss in this paper.

The reason for the THz gap is the THz photons (1THz frequency = 4.2 meV photon energy) have an energy much smaller than thermal energy at room temperature (25.2 meV) making it difficult to implement THz lasers and light-emitting diodes (LEDs). On the other hand, even though the cutoff frequency, $f_T = v_s/(2\pi L)$, of modern transistors ideally reaches 1 THz (see Fig. 3), in fact, it is smaller because of parasitics. (Here v_s is the effective electron velocity in the channel, and L is the channel length.) The minimum feature size of modern Si CMOS is 3 nm (used, for example, in iPhone 15 Pro). However, the effective channel is closer to 20 nm.

As explained in introductory quantum mechanics courses, electrons have particle-like and wave-like properties. However, in modern short-channel transistors, electrons often behave like an electronic fluid. Such electronic fluid exists when the electron–electron collision time, τ_{ee}, is much shorter than all other relevant characteristic time scales, such as the momentum relaxation time, τ_m. Electronic fluid supports waves of electron density. Mathematics describing these waves is the same as describing water or sound waves [3]. When the device length becomes smaller than the mean

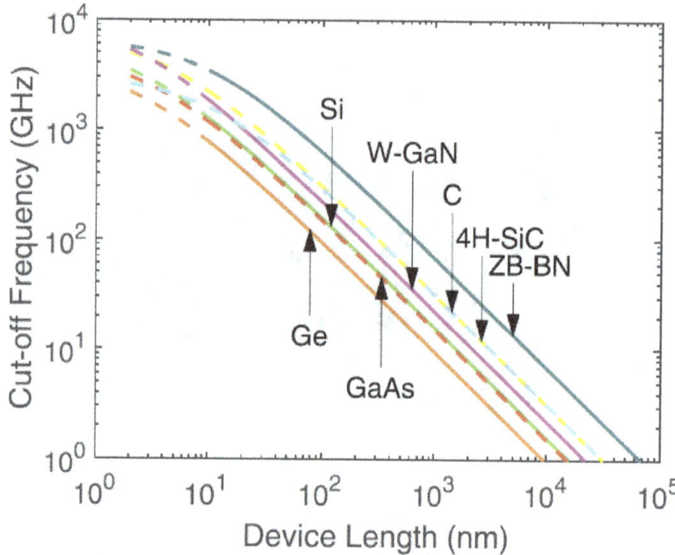

Fig. 3.   FET cutoff frequency limit versus channel length for different material systems (from [2]).

free path for collisions with impurities and lattice vibration, the electron transport becomes ballistic or quasi-ballistic [4, 5]. These effects completely change the physics of electron transport in short-channel devices [6–14]. An electron mean free path in Si at room temperature is in the range from 20 nm to 50 nm [10] (see Fig. 4).

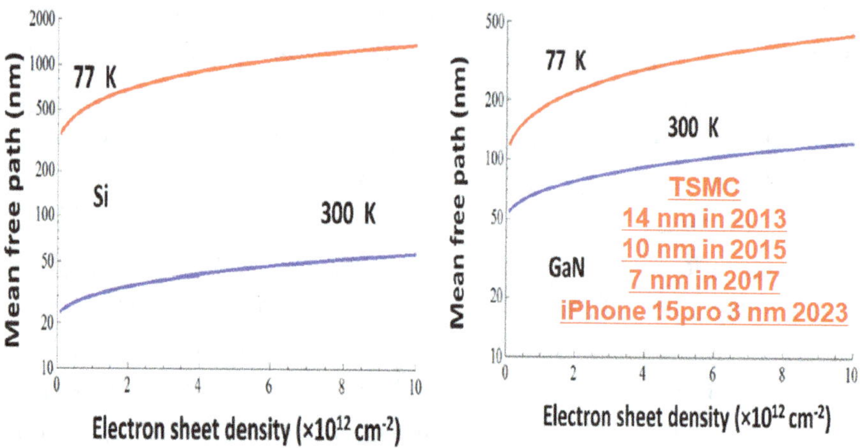

Fig. 4.   Mean free path at room temperature and 77 K [10].

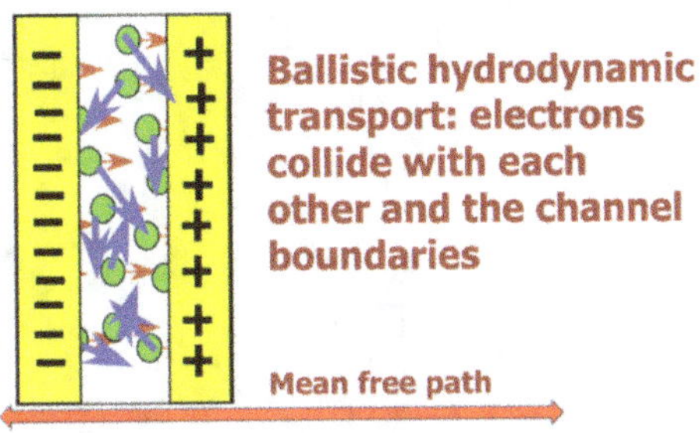

Fig. 5.   Ballistic hydrodynamic transport: electrons collide with each other and the channel boundaries [10].

One of the consequences of ballistic transport is that electrons lose their momentum by scattering when hitting the contacts, as schematically shown in Fig. 5. As a result, the effective (measured) mobility (called ballistic mobility [11, 12]) decreases proportionally to the length (see Fig. 6).

Ballistic transport and fast propagation of plasma waves in short-channel FETs dramatically change the expectations of the FET switching time and the modulation frequency. Launching, detecting, and generating plasma waves, i.e., operating short-channel transistors in plasmonic regimes allows for reaching much higher frequencies of operation (see Figs. 7–10).

As seen from Fig. 7, in plasmonic modes, FETs could operate about an order of magnitude higher than in the conventional transit modes. Response times of short-channel Si CMOS and compound semiconductor FETs could reach the subpicosecond range (see Figs. 8 and 9). (Such short-channel FETs operating in plasmonic, ballistic, or quasi-ballistic regimes are often called TeraFETs.) As seen from these figures, ballistic mobility is dominant in high mobility-short channel devices leading to an increase in the response time (and a commensurate decrease of the modulation frequency). For high-mobility, short-channel structures, the response increases with mobility because of plasmonic resonant ringing. At very high mobilities, like in high-quality graphene, even at room temperature, the response time becomes nearly independent of

                    M. Shur

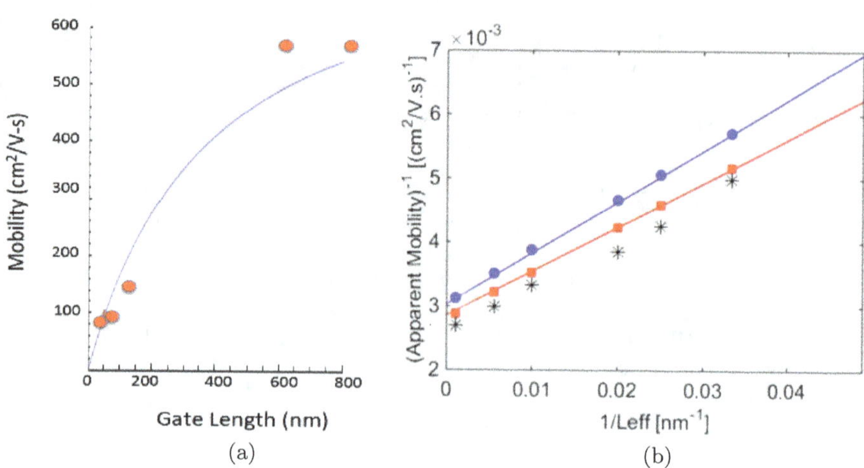

Fig. 6.   Measured field effect mobility in Si (a) from [13] and (b) inverse mobility versus inverse length from [14]. Squares and circles correspond to Vgs = 0.5 and 0.8 V, respectively (Ns = 6×10^{12} and 10^{13} cm^{-2}). Stars are from mobilities obtained by fitting to the experimental data as described in [15].

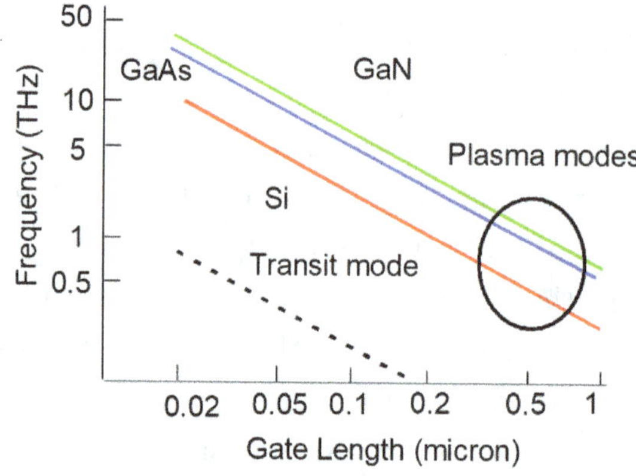

Fig. 7.   Transit and plasmonic modes of operation [16].

mobility and is dominated by the viscosity of the electronic fluid (see Fig. 9).

Accounting for this counter-intuitive device physics is crucial for device simulation, design, parameter extraction, and model validation [17, 18]. It could bring down an astronomical cost of

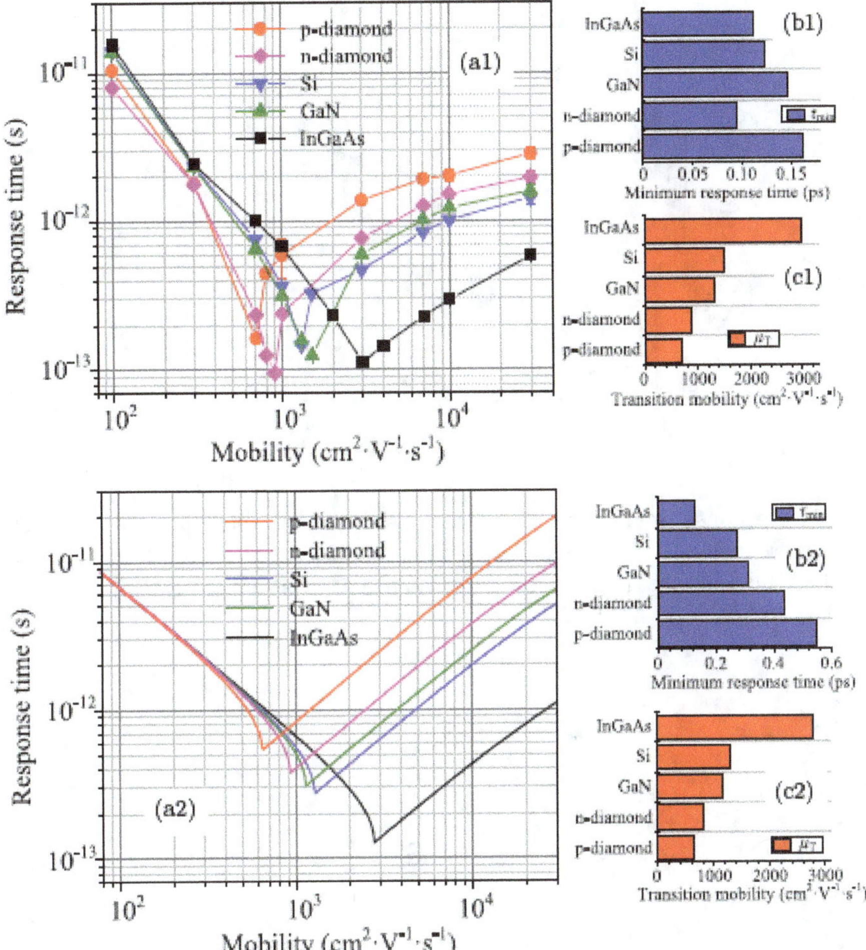

Fig. 8. Ultimate response times of TeraFETs for five materials versus carrier mobility (a1) and (a2), minimum response time (b1) and (b2), and transition mobility (c1) and (c2). (a1)–(c1) show the simulation data and (a2)–(c2) present the analytical results. The gate voltage swing U_0 is set at 0.1 V, the gated channel length L is 130 nm, and the pulse width of the incoming radiation is 5×10^{-14} s, under room temperature ($T = 300$ K) [19].

the design of modern VLSI chips reaching U\$1.5 billion for VLSI at the 3 nm technology node (like those used in iPhone 15Pro). It is even more important for developing new integrated circuits operating in the THz range of frequencies with near near-term

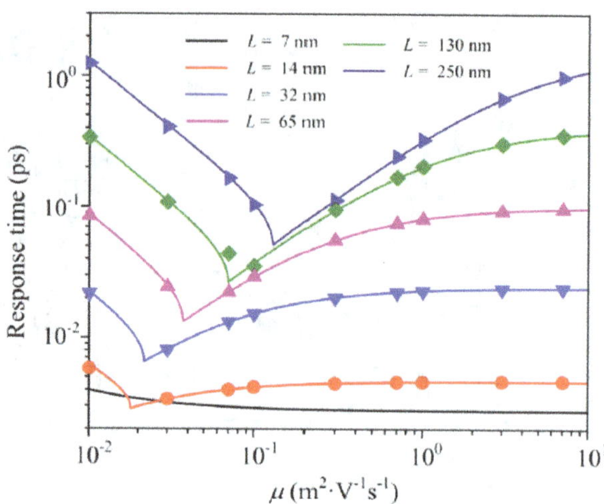

Fig. 9.   Response time versus mobility μ for bilayer graphene plasmonic FETs with different channel lengths. The symbols represent the simulation values, and the solid lines are the analytic curves. The gate voltage swing 2 V, viscosity $\nu = 0.034\,\mathrm{m^2/s}$, $T = 300\,\mathrm{K}$ [20].

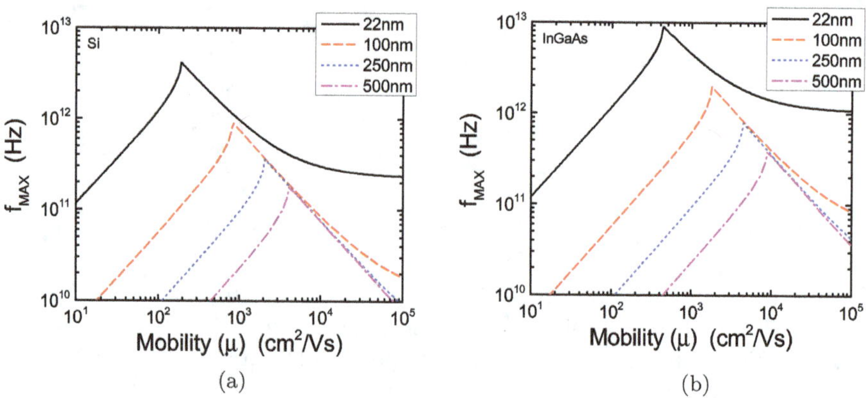

Fig. 10.   Maximum modulation frequency for Si and InGaAs FETs [21].

goal of developing VLSI operating in the 300 GHz range (the Earth atmosphere window for 6G communication.

These dependencies of the response times lead to inverse dependencies of the maximum modulation frequency (see Fig. 10).

2. Terahertz Sensors and Sources

THz radiation impinging on a THz sensor induces a response (a voltage varying at the THz frequency $\omega = 2\pi f$):

$$U = U_o \cos \omega t. \tag{4.1}$$

The induced response (such as a current, I, flowing in the sensor circuit) is nonlinear. At low excitation intensities, this response can be approximated as

$$I = \sigma U_o \cos \omega t + \alpha U_o^2 \cos^2 \omega t = \sigma U_o \cos \omega t + \tfrac{1}{2}\alpha U_o^2(1 + \cos 2\omega t). \tag{4.2}$$

As seen from Eq. (4.2), the rectified response is

$$I_{ph} = \alpha U_o^2/2. \tag{4.3}$$

If the intensity of the THz radiation is modulated at some frequency ω_m the response could be modulated too (provided that the characteristic response time, $\tau_R < 1/\omega_m$. Examples of such THz detectors are Schottky diodes and TeraFETs.

A THz bolometer senses the temperature change caused by the absorbed energy of the THz radiation that affects the detector temperature or the temperature of current carriers (electrons or holes or both) in hot-electron bolometers. The change in temperature leads to a change in the output current.

Another THz detector is a Golay cell using THz radiation absorption in a material enclosed in a gas-filled chamber. The increasing gas temperature leads to gas expansion moving a membrane illuminated by an LED. A photodiode detects the reflected light signal that is proportional to the absorbed THz radiation.

Photodetector responsivity is the ratio of the detector's electrical output (current, I_{ph}, or voltage, V_{ph}) over the impinging radiation power, P:

$$R_I = I_{ph}/P(A/W); \quad R_V = V_{ph}/P(V/W). \tag{4.4}$$

If each photon of the incoming radiation creates an electron–hole pair that contributes to the output photodetector current, the current responsivity would be equal to

$$R_{\text{IMAX}} = \frac{q}{\hbar\omega} = \frac{q\lambda}{2\pi\hbar c} \approx \frac{\lambda(\mu m)}{1.24}. \tag{4.5}$$

Here q is the electronic charge, $\hbar$ is the Planck constant, c is the speed of light, and λ is the wavelength. A more realistic equation for R_I should account for the quantum efficiency, η

$$R_I = \eta R_{\text{IMAX}} = \eta\frac{q}{\hbar\omega} = \eta\frac{q\lambda}{2\pi\hbar c} \approx \eta\frac{\lambda(\mu m)}{1.24}. \tag{4.6}$$

An unavoidable noise signal $i_n(t)$ is superimposed on the sensor response

$$I_{ph} = I_{\text{pho}} + i_n(t). \tag{4.7}$$

Noise is characterized by the power spectral density (PSD). PSD is measured in V^2/Hz or in $V/rt\ Hz$ or in A^2/Hz. Here rt Hz means "square root of Hertz". The overall noise is the integral of the spectral distribution over the bandwidth. Figure 11 shows typical noise spectra for different noise mechanisms.

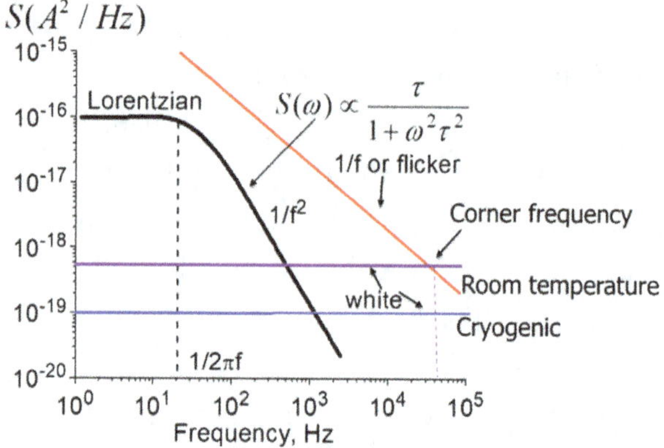

Fig. 11.   Typical noise spectra and corner frequency.

For the white noise, the noise voltage is given by

$$\langle V^2 \rangle = 4k_B T R \times BW. \tag{4.8}$$

Here k_B is the Boltzmann constant, T is temperature, R is the detector resistance and BW is the bandwidth.

Other noise mechanisms (not represented in Fig. 12) include shot noise (also called Schottky noise or Poisson noise) caused by the fluctuations in the number of current carriers

$$\overline{i_{\mathrm{shot}}} = (2q\overline{I} \times BW)^{1/2}. \tag{4.9}$$

Here q is the electronic charge, and I is the average current. Hence, the low bound of the white noise current (ignoring important or even dominant contributions from the frequency-dependent noise) is

$$i_{\mathrm{noise}} = \sqrt{(4k_B T/R + 2q\overline{I}) \times BW}. \tag{4.10}$$

The frequency-dependent noise mechanisms include recombination-generation noise caused by the fluctuations of the generation and recombination processes in a photodetector and having the Lorentzian power spectral distribution, flicker $(1/f)$ noise (see Fig. 11), and random telegraph noise (also called popcorn noise or burst noise. The burst noise is caused by step-like transitions between two voltage or current levels at random times. The flicker $(1/f)$ noise in photodetectors is linked to mobility or carrier concentration fluctuations. It is typical for a large variety of different phenomena ranging from photodetectors to fluctuations of blood sugar, transparency of the atmosphere, oscillations of Earth's crust, overflows of the Nile River for 2800 years, cordial rhythms, the potential of biological cells, the quantity of insulin in human blood, pain relief efficacy, traffic, and even musical patterns.

Table 1 presents a summary of different noise mechanisms.

The overall noise characteristic of a sensor is the Noise Equivalent Power (NEP). NEP (measured in $W/\sqrt{Hz}$) is the input signal power that results in a signal-to-noise ratio (S/R) of 1 in

Table 1.   Mechanisms of noise.

	Devices affected	Mechanism	Depends on
	White (frequency-independent) noise		
Thermal	All	Thermal motion	Temperature, resistance
Shot			
	Frequency-dependent noise		
Flicker ($1/f$)	All materials,	In photodiodes: mobility	Current, material
	devices, and many phenomena	or concentration fluctuations	quality, surface, temperature
Generation/	Photodetectors	Generation/ recombination	Temperature, bias
recombination		rate fluctuations	
RTF (burst, popcorn)	MOSFETs	Electron capture/ emission	Temperature, strain, frequency

a $1\,\mathrm{Hz}$ output bandwidth. The NEP decreases inversely to the square root of the averaging time t_{average}. According to the Nyquist theorem

$$t_{\text{average}} = 1/(2BW). \qquad (4.11)$$

NEP could be interpreted as the minimum power that could be detected by a photodetector. It could be defined as $\mathrm{NEP} = S_I/R_I$ or $\mathrm{NEP} = S_V/R_V$. The related characteristic is specific detectivity.

$$D* = \sqrt{A}/NEP = \sqrt{A}R_V/S_V. \qquad (4.12)$$

Here A is the detector area. Detectivity is measured in $m\sqrt{Hz}/W$ $m\sqrt{Hz}/W$ or in Jones ($cm\sqrt{Hz}/W$). Table 2 lists the typical characteristics of different THz photodetectors.

Table 3 lists THz sources, which are the bottleneck of compact electronic THz technology. TeraFETs operating in plasmonic regimes have the potential to revolutionize this technology. They

Table 2.   THz photodetectors.

Device	R_V (V/W)	NEP	T (K)	f (THz)	Response frequency (Hz)
Golay cell	500–2500	~100	300	0.05–50	~100 Hz
Bolometers (uncooled)	1000–2500	5–300	300	0.1–30	100 Hz–1 kHz
Bolometers (cooled)	10^5	0.25	1.6–4.2	0.2–20	100 Hz–1 kHz
Schottky diodes	100–3000	3–300	300	0.05–10	10^{10}–5×10^{10}
Si MOSFETs	50–2000	15–5000	300	0.2–5	10^{10}–5×10^{10}
AlGaN/GaN HEMTs	150000	0.5–30	300	0.2–5	10^{10}–5×10^{10}
GaAs HEMTs	50–100	100	300	0.2–5	10^{10}–5×10^{10}
InP HEMTs	20000	0.5	300	0.2–5	10^{10}–5×10^{10}
Graphene bolometers	10–700	30–2000	300	0.13–3	10^{10}–5×10^{10}

Table 3.   THz sources.

THz source	Power range	Frequency (THz)	T(K)	Tunability	Output
Photoconductive antenna	μW	0.1–10	300	—	Pulsed
THz laser	W	0.3–10	77	Discrete lines	CW/pulsed
Quantum cascade laser	mW	2–10	Cryogenic	10 GHz	CW/pulsed
Multipliers (Schottky diodes)	mW–μW decreasing with frequency	0.1–10	300	15%	CW
Backward wave oscillator	mW	0.1–1.5	300	20%	CW/pulsed
Free electron lasers	100 MW	0.2–30	300	Varied	Pulsed
TeraFET	nW	0.3–10	77–300	50%	CW

now produce very little power but using phase-matched arrays could increase this power to the mW range.

TeraFET array and plasmonic crystal technology [22–31], which is now under development, has the best chance to support the transformational shift to 6G communications [32, 33] simultaneously dramatically reducing the cost and complexity of THz sensing applications.

3. Applications of Terahertz Sensing

Figure 12 (from [34]) maps different THz applications within the THz spectrum range. The killer application is wireless 6G communication in the 300 GHz band. After 2028, 5G communications (operating in the 30 GHz band or below) will not be able to cope with the global data traffic and the transition to 6G communications will become unavoidable. THz sensing applications include industrial quality control and defect identification (see Fig. 13 demonstrating unique capabilities of the THz technology) to gas detection (see Table 4).

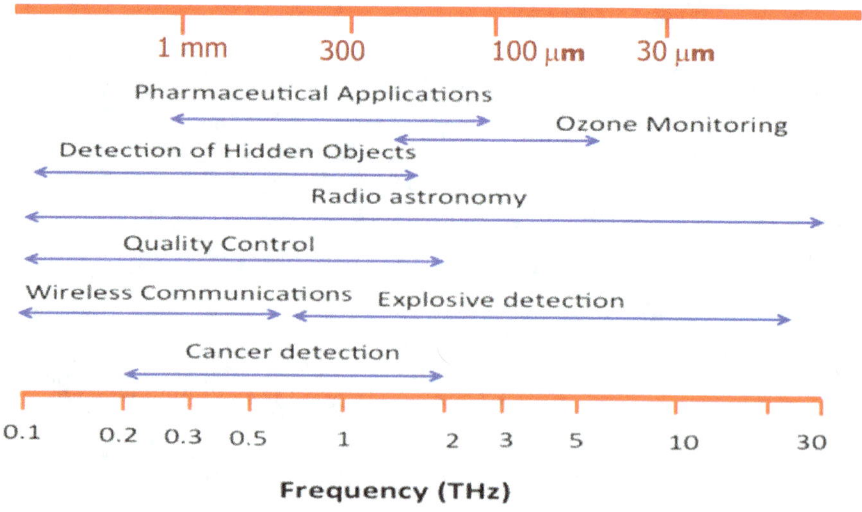

Fig. 12.   THz applications within the THz spectrum range.

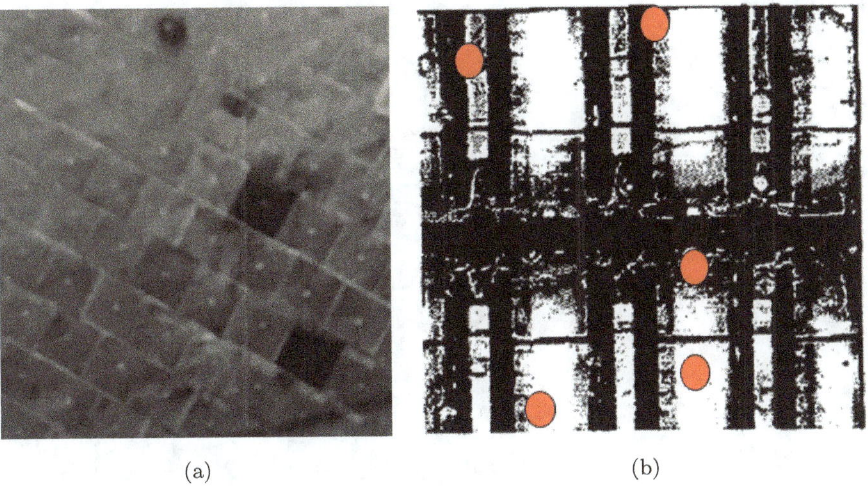

(a)   (b)

Fig. 13.   Testing faults in space shuttle tiles. Red circles marked the detected faults.

Table 4.   Gas sensing using THz radiation.

THz range (THz)	Compound
0.1–1.8	Trichloromethane
0.1–2.0	Carbon monoxide, nitrogen monoxide, carbendazim
0.1191–1195	$(HCN)_2$
0.2–1.5	Methane, ethane
0.2–2.6	Sulfur dioxide, hydrogen sulfide
0.21–0.27	Hydrocyanic acid, acetonitrile
0.2468–0.2612	Fluoromethane
0.3–10	Carbon disulfide
0.44–0.49	Nitrogen monoxide
0.5–2	Methomyl
0.5–6.2	Butane

Molecular vibrations, bond vibrations, hydrogen bond oscillations, and phonons all map into the THz range (see Fig. 14) enabling numerous biomedical applications (see Fig. 15) including cancer detection (see Figs. 15–18) and, possibly, even cancer treatment. THz radiation is well suited for medical and biomedical applications because THz absorption or reflection from biological

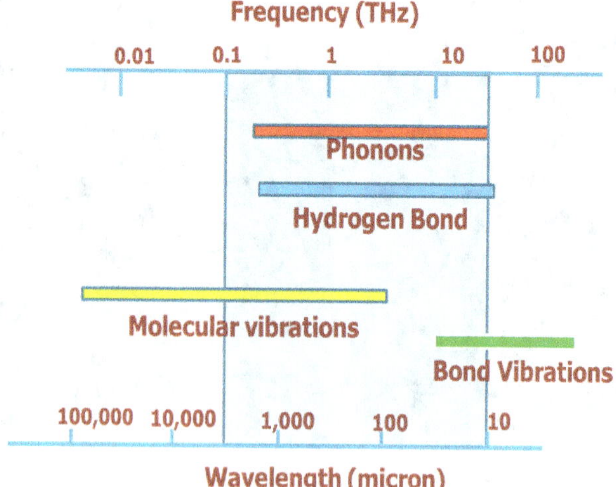

Fig. 14.   THz range (0.1–30 THz) for organic and chemical matter detection [35].

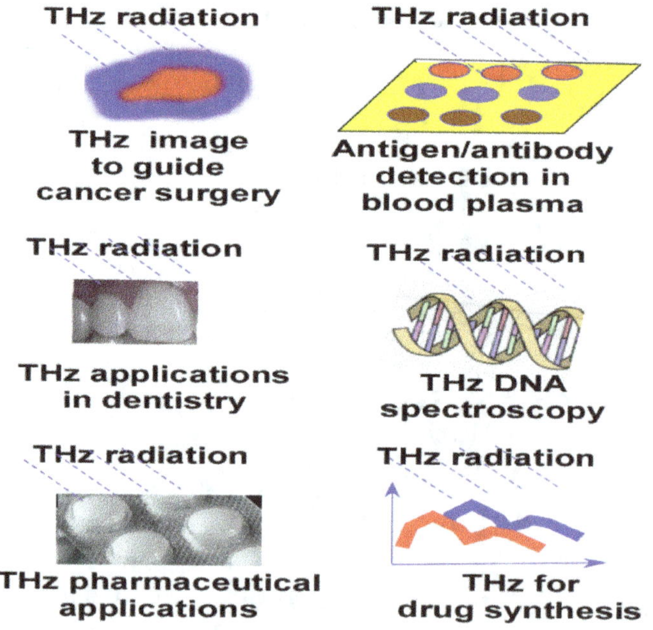

Fig. 15.   Biomedical applications of THz sensing [36].

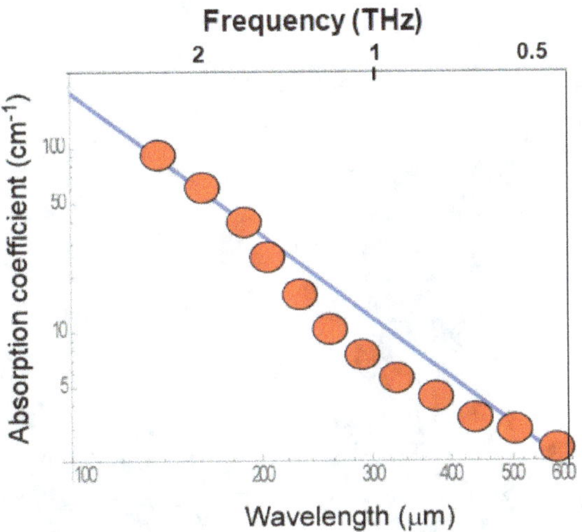

Fig. 16. THz absorption in water interpolated $\alpha = (\lambda_o/\lambda)^m$. Here $\lambda_o = 786\,\mu$m, $m = 2.57$ [35].

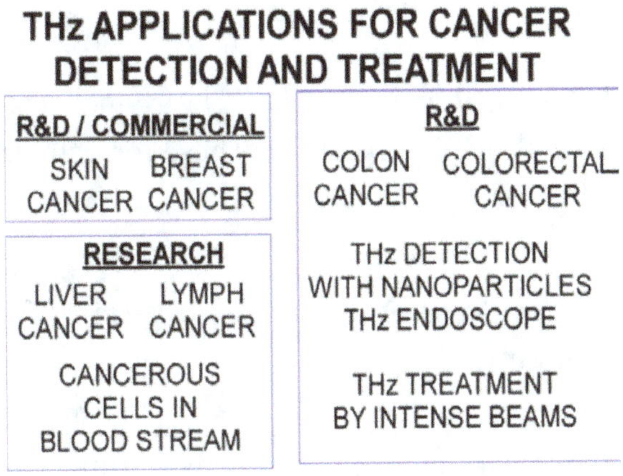

THz APPLICATIONS FOR CANCER DETECTION AND TREATMENT

Fig. 17. State-of-the-art of THz cancer detection [36].

tissues yields their detailed spectroscopic signatures without any ionization hazard. The terahertz time-domain spectroscopy (THz-TDS) systems detect the optical properties of tissues with different water content at terahertz frequencies. The THz absorption

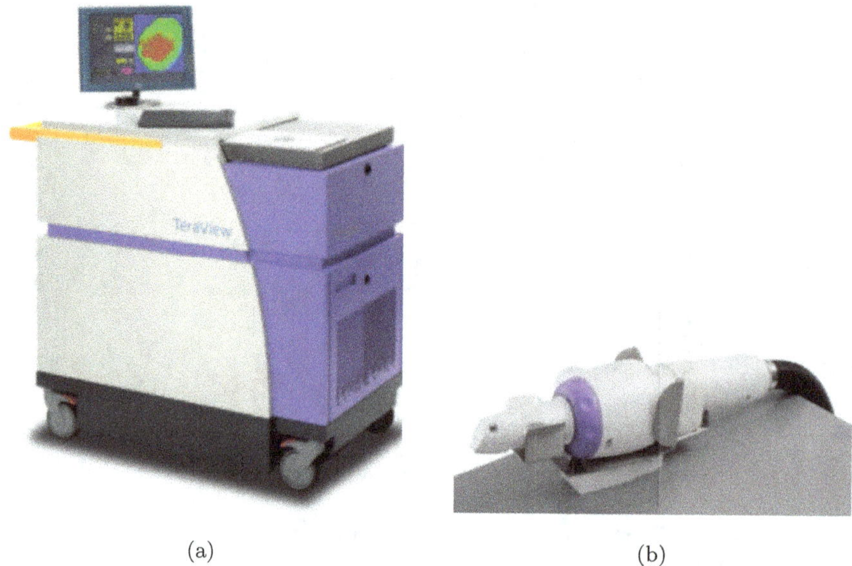

Fig. 18.   TeraView System for breast cancer imaging (a) and handheld THz scanner for cancer detection (b) [37].

coefficient and the refractive index at THz frequencies of hydrated biological tissues are much larger than those for dehydrated tissues (see Fig. 14).

4.   THz Market, Companies, and Countries Working on THz Sensing

THz market is expected to grow at the Compound Annual Growth rate of 26.4% reaching approximately \$3000 million by 2030. My expectations (schematically shown in Fig. 19) are higher due to the expected deployment of 6G communication technology operating in the 300 GHz band in the 2026–2028-time frame.

An incomplete but representative list of companies working on THz sensing ranging from large to small is provided in Table 5.

The interest of different countries in THz technology could be indirectly gauged using the number of patent applications for 6G technologies — the technologies that will drive all other THz sensing applications as well (see Fig. 20).

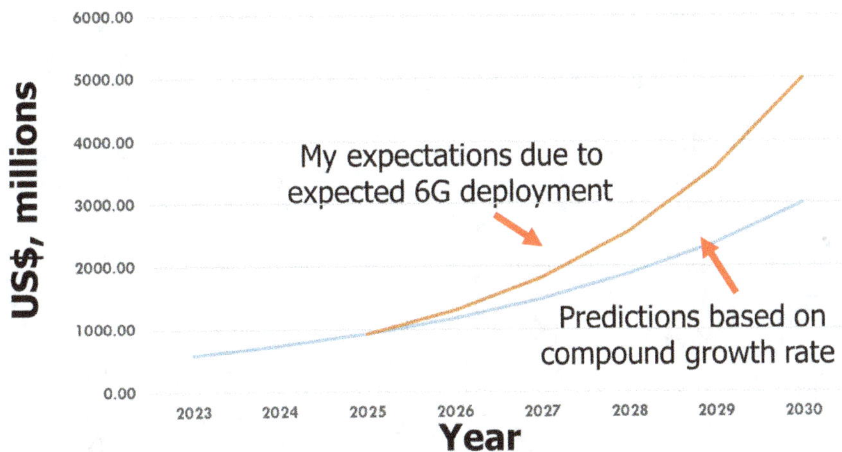

Fig. 19. THz technology market expectations (data for compound annual growth rate (from [37]).

Table 5.   Terahertz companies.

Company	Founded	Web site	Country
GlaxoSmith-Kline	1830	http://www.gsk.com	UK
Thales	1918	http://www.thalesgroup.com	France
Physical Sciences	1973	http://www.psicorp.com/	USA
TeraView	2001	http://www.teraview.com/	UK
TiHive	2017	https://www.tihive.com/	France
Electronics of the Future	2017	http://www,electronicsoffuture.com	USA
Virginia Diodes	1996	http://www.VADiodes.com	USA
TeraSense	2009	http://terasense.com/	USA
LUNA	1990	https://lunainc.com/	USA
Teravil	2006	https://www.teravil.lt/t-spec.php	Japan
Toptica Photonics	1998	https://www.toptica.com/	USA
HUBNER Photonics	1946	https://hubner-photonics.com/	USA
Tydex	1994	www.Tydexoptics.com	Russia
THORLABS	1989	www.Thorlabs.com	USA
Hamamatsu	1953	https://www.hamamatsu.com/us/en.html	Japan

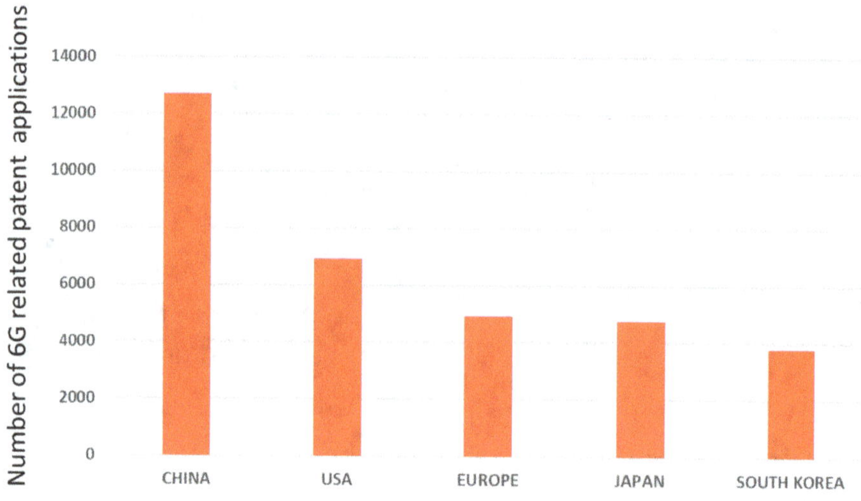

Fig. 20.   6G technology patent applications by country (data from [38]).

5. Conclusion

In conclusion, THz sensing has found numerous applications ranging from 6G communications to industrial controls, chemical analysis, biomedical applications, cancer diagnostics, and, possibly, even treatment. These and other applications are hampered by the high cost of THz components. Plasma electronics, TeraFETs, especially TeraFET arrays — plasmonic crystals — have the potential of changing it in the near future. Transformational progress in THz sensing will be enabled by the pent-up demand for 6G communications in the 300 GHz band required to meet the growing demand for rapidly growing data traffic that can no longer be handled by 5G after 2028. This transformation will require Si/GaN hetero-technology integration and sensing both the intensity and phase of THz radiation.

ORCID

Michael Shur Ssali https://orcid.org/0000-0003-0976-6232

References

1. https://www.chemistryworld.com/opinion/argands-lamp/3009536.article, accessed Sep-tember 16, 2023.
2. J. Chilleri, P. Siddiqua, M. S. Shur and S. K. O'Leary, Cubic boron nitride as a material for future electron device applications: *A comparative analysis*, *Appl. Phys. Lett.* (*Invited*), **120**, 122105 (2022), doi:10.1063/5.0084360.
3. M. Dyakonov and M. S. Shur, Shallow water analogy for a ballistic field effect transistor. New mechanism of plasma wave generation by DC current, *Phys. Rev. Lett.*, **71**(15), 2465–2468 (1993).
4. M. S. Shur and L. F. Eastman, Ballistic transport in semiconductors at low-temperatures for low power high-speed logic, *IEEE Trans. Electron Devices*, ED-**26**(11), 1677–1683 (1979).
5. M. S. Shur, Ballistic and collision dominant transport in a short semiconductor diode, *Proc. IEDM, Washington, DC*, 1980, pp. 618–621.
6. M. S. Shur, Ballistic transport in semiconductor with collisions, *IEEE Trans. Electron Devices*, ED-**28**(10), 1120–1130 (1981).
7. K. Lee and M. S. Shur, Impedance of thin semiconductor film. *J. Appl. Phys.*, **54**(7), 4028–4034 (1983).
8. A. van der Ziel, M. S. Shur, K. Lee, T. H. Chen and K. Amberiadis, Carrier distribution and low-field resistance in short n+-n-n+ structures, *IEEE Trans. Electron Devices*, ED-**30**(2), 128–137 (1983).
9. J. Xu and M. S. Shur, Ballistic transport in hot electron transistors, *J. Appl. Phys.*, **62**(9), 3816–3820 (1987).
10. M. S. Shur, Terahertz plasmonic technology, *IEEE Sens. J.* (*Invited*), **21**(11), 12752–12764 (2021), doi:10.1109/JSEN.2020.3022809.
11. A. A. Kastalsky and M. S. Shur, Conductance of small semiconductor devices, *Solid State Commun.*, **39**(6), 715–718 (1981).
12. A. P. Dmitriev and M. S. Shur, Ballistic admittance: Periodic variation with frequency, *Appl. Phys. Lett.*, **89**, 142102 (2006).
13. W. Knap, F. Teppe, Y. Meziani, N. Dyakonova, J. Lusakowski, F. Bouef, T. Skotnicki, D. Maude, S. Rumyantsev and M. S. Shur, Plasma wave detection of millimeter wave radiation by silicon field effect transistors, *Appl. Phys. Lett.*, **85**(4), 675–677 (2004) (also re-published in Virtual J. Nanoscale Sci. Technol., 6 August 2004).
14. D. Antoniadis, On apparent electron mobility in Si nMOSFETs from diffusive to ballistic regime, *IEEE Trans. Electron Devices*, **63**(7), 2650–2656 (2016), doi:10.1109/TED.2016.2562739.
15. F. Ferdousi, R. Rios and K. J. Kuhn, Improved MOSFET characterization technique for single channel length, scaled transistors, *Solid-State Electron.*, **104**, 44–46 (2015).
16. V. Ryzhii and M. S. Shur, Plasma wave electronics devices, ISDRS Digest, WP7-07-10, pp. 200–201, Washington, DC (2003).

17. M. Shur, Terahertz compact SPICE model, 2016 MIXDES — 23rd International Conference Mixed Design of Integrated Circuits and Systems, Lodz, Poland, 2016, pp. 27–31, doi:10.1109/MIXDES.2016.7529694.S.

18. A. Gutin, T. Ytterdal, V. Kachorovskii, A. Muraviev and M. Shur, THz SPICE for modeling detectors and non-quadratic response at large input signal, *IEEE Sens. J.*, **13**(1), 52–62 (2013), doi: 10.1109/JSEN.2012.2226647.

19. Y. Zhang and M. S. Shur, p-Diamond, Si, GaN and InGaAs TeraFETs, *IEEE Trans. Electron Devices*, **67**(11), 4858–4865 (2020), doi:10.1109/TED.2020.3027530.

20. Y. Zhang and M. S. Shur, Collision dominated, ballistic and viscous regimes of terahertz plasmonic detection by graphene, *J. Appl. Phys.* **129**, 053102 (2021), doi:10.1063/5.0038775.

21. G. Rupper, S. Rudin and M. Shur, Response of plasmonic terahertz detectors to amplitude modulated signals, *Solid State Electron.*, **111**, 76–79 (2015).

22. M. Shur and R. Gaska, Device and method for managing radiation, US Patent Application 20060081889, April 20, 2006, US patent 7638817 issued December 29, 2009; US Patent 8,053,271.

23. M. Shur, V. Ryzhii and R. Gaska, Method of radiation generation and manipulation, US Patent 7,619,263, November 17 (2009); US Patent 7,955,882.

24. A. Rogalski and F. Sizov, Terahertz detectors and focal plane arrays, *Opto-Electron. Rev.*, **19**(3), 346–404 (2011), doi:10.2478/s11772-011-0033-3.

25. K. Song, D. Kim, J. Kim, J. Yoo, W. Keum and J.-S. Rieh, A scalable 300-GHz multichip stitched CMOS detector array, *IEEE Trans. Microwave Theory Tech.*, **70**(3), 1797–1809 (2022).

26. W. Knap *et al.*, Terahertz imaging with arrays of plasma field effect transistors detectors, 2016 41st International Conference on Infrared, Millimeter and Terahertz waves (IRMMW-THz), Copenhagen, Denmark, 2016, pp. 1–2, doi:10.1109/IRMMW-THz.2016.7758828.

27. R. Aizin, J. Mikalopas and M. Shur, Plasmonic instabilities in two-dimensional electron channels of variable width, *Phys. Rev. B*, **101**, 245404 (2020).

28. A. S. Petrov, D. Svintsov, V. Ryzhii and M. Shur, Amplified-reflection plasmon instabilities in grating-gate plasmonic crystals, *Phys. Rev. B*, **95**, 045405 (2017).

29. T. Otsuji, Y. M. Meziani, T. Nishimura, T. Suemitsu, W. Knap, E. Sano, T. Asano and V. V. Popov, Emission of terahertz radiation from dual grating gate plasmon-resonant emitters fabricated with InGaP/InGaAs/GaAs material systems, *J. Phys. Condens. Matter*, **20**, 384206 (2008) and references therein.

30. G. R. Aizin, J. Mikalopas and M. Shur, Current driven "plasmonic boom" instability in gated periodic ballistic nanostructures, *Phys. Rev. B*, **93**(19), 195315 (2016).

31. G. R. Aizin, J. Mikalopas and M. Shur, Current driven Dyakonov–Shur instability in ballistic nanostructures with a stub, *Phys. Rev. Appl.*, **10**, 064018 (2018).

32. M. Z. Chowdhury, M. Shahjalal, S. Ahmed and Y. M. Jang, 6G wireless communication systems: Applications, requirements, technologies, challenges and research directions, *IEEE Open J. Commun. Soc.*, **1**, 957–975 (2020), doi:10.1109/OJCOMS.2020.3010270.

33. M. Shur, G. Aizin, T. Otsuji and V. Ryzhii, Plasmonic field-effect transistors (TeraFETs) for 6G communications, Sensors, **21**, 7907 (2021).

34. T. Otsuji and M. S. Shur, Terahertz plasmonics: Good results and great expectations, *IEEE Micro. Mag.* **15**(7), 43–50 (2014), doi:10.1109/MMM.2014.2355712.

35. M. Shur, Subterahertz and terahertz sensing of biological objects and chemical agents, Proc. SPIE 10531, Terahertz, RF, Millimeter and Submillimeter-Wave Technology and Applications XI, 1053108, 23 February 2018, doi:10.1117/12.2288855.

36. M. Shur and X. Liu, Keynote, Proc. SPIE 11975, Advances in Terahertz Biomedical Imaging and Spectroscopy, 1197502, 3 March 2022, doi:10.1117/12.2604800.

37. https://www.photonics.com/Products/Terahertz_spectrometer/pr33589, accessed October 21, 2023.

38. https://www.lexology.com/library/detail.aspx?g=ecf6c614-ec71-4b2b-9a46-a7781de1210c, accessed October 20, 2023.

Chapter 5

Phase- and Angle-Sensitive Terahertz Hot-Electron Bolometric Plasmonic Detectors Based on Fets with Graphene Channel and Composite H-BN/Black-P/H-BN Gate Layer[#]

Victor Ryzhii[*,†,‖], Chao Tang[†], Taiichi Otsuji[†], Michael S. Shur[‡,**], Maxim Ryzhii[§,††] and Vladimir Mitin[¶,‡‡]

[†]*Research Institute for Electrical Communication, Tohoku University, Sendai, Miyagi 980-8577, Japan*
[‡]*Department of Electrical, Computer, and Systems Engineering, Rensselaer Polytechnic Institute, Troy, New York 12180, USA*
[§]*Department of Computer Science and Engineering, University of Aizu, Aizu-Wakamatsu, Fukushima 965-8580, Japan*
[¶]*Department of Electrical Engineering, University at Buffalo, SUNY, Buffalo, New York 14260, USA*
[‖]*vryzhii@gmail.com*
[**]*shurm@rpi.edu*
[††]*m-ryzhii@u-aizu.ac.jp*
[‡‡]*vmitin@buffalo.edu*

---

[*]Corresponding author.

[#]This chapter appeared previously on the International Journal of High Speed Electronics and Systems. To cite this chapter, please cite the original article as the following: V. Ryzhii, C. Tang, T. Otsuji, M. S. Shur, M. Ryzhii and V. Mitin, *Int. J. High Speed Electron. Syst.*, **33**, 2440023 (2024), doi:10.1142/S0129156424400238.

In this paper, we propose and analyze the terahertz (THz) bolometric vector detectors based on the graphene-channel field-effect transistors (GC-FET) with the black-P gate barrier layer or with the composite b-BN/black-P/b-BN gate layer. The phase difference between the signal received by the FET source and drain substantially affects the plasmonic resonances. This results in a resonant variation of the detector response on the incoming THz signal phase shift and the THz radiation angle of incidence.

Keywords: Graphene; terahertz radiation; hot-electron bolometer; field-effect transistor; phase sensitivity; angle sensitivity.

1. Introduction

The specific properties, in particular, the band alignment of graphene and the black-P and black-As (or black-AsP) moderately thick layers [1–4] enable the operation of different devices using the thermionic electron or hole emission from the graphene channel (GC) via the gate barrier layer (BL). Recently [5,6], we predicted that the hot-electron bolometric detectors based on the GC-FETs with the GC and the black-AsP BL can exhibit fairly high characteristics, for example, responsivity. The operation of such detectors is associated with the electron heating by the impinging terahertz (THz) radiation stimulating the thermionic electron emission from the GC via the black-AsP BL into the metal gate (MG). The excitation of the plasmonic oscillations in the gated GC enables the resonant increase in the electron heating reinforcing thermionic emission. Plasmonic resonances can provide elevated detector performance [5,6].

The strength of the plasmonic resonances depends on the electron collision frequency ν (determining the plasmonic oscillation damping), i.e., on the electron mobility in the GC. Since the GC contacting the black-AsP exhibits a moderate room temperature mobility [7] and not particularly low electron collision frequency, the GC-FETs with the composite h-BN/black-AsP/h-BN gate BLs could be more promising. Indeed, the GC-FETs with the GC main part encapsulated in the h-BN operating at room temperature can provide a fairly pronounced resonant response due to high electron mobility and, therefore, a relatively small ν [8, 9]. This challenge can be overcome in similar GC-FET bolometric detectors but with the composite incorporating a relatively narrow black-AsP placed between the h-BN side sections. In these detectors, the black-AsP section of the BL serves as the window for the thermionic electron current, while the rest of the h-BN can enable a much weaker electron scattering in the GC. As a result, such GC-FETs can demonstrate a more pronounced resonant response and, hence, elevated responsivity [10].

One of the GC-FET detector's feature is a strong influence of the signal electric field spatial distribution along the GC on the plasmonic response. This implies that this distribution can be varied by changing the phase difference between the signal voltages created by the antenna system at the GC-FET side contacts (source and drain).

In this paper, we consider the GC-FET bolometric detector structures akin to those proposed and evaluated previously [5,6,10], but focusing on their sensitivity to the THz radiation phase and incident angle. Such a property of THz detectors can be useful for different application (see [11]) and the references therein). For the definiteness, we consider the GC-FET bolometric detectors with the GC and the composite h-BN/black-P/h-BN gate BL.

2.   Device Structure

Figure 1(a) shows the GC-FET structures under consideration. The GC-FET comprises the GC-sandwiched between the h-BN bottom layer (substrate) and the BL consisting of the lateral h-BN/black-P/h-BN composite. Figure 1(b) shows the GC-FET band diagram at the BL central section (BL window, $|x| \leq L_C$). Here $2L$ is the length of the GC (approximately equal to the MG length), $2L_C$

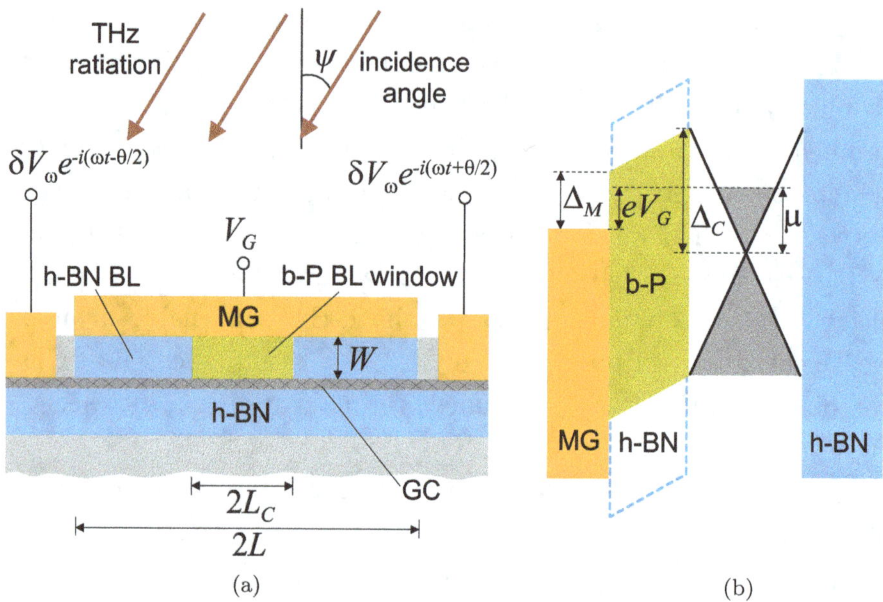

Fig. 1. Schematic view of the structure of (a) GC-FET bolometric detector structure with the composite h-BN/black-P/h-BN BL, (b) band diagram in the central black-P section of the BL (in the BL window; dashed line corresponds to shape of the h-BN BL in the side regions).

is the length of the black-AsP section, and the axis x is directed in the GC plane between the side contacts. As an alternative, the WSe_2 side sections of the BL and/or the WSe_2 bottom layer can be used due to the high room temperature mobility [12] (and, hence, a low electron collision frequency) of the GC containing such a BL.

The GC channel is n-doped, so that the electron Fermi energy $\mu \simeq \hbar v_W \sqrt{\pi \Sigma_D} \gg T_0$, where Σ_D is the donor density in the GC or in the neighboring layers, T_0 is the lattice temperature, $\hbar$ is the Planck constant and $v_W \simeq 10^8$ cm/s is the electron velocity in graphene. The material of the MG is chosen to provide the condition $\Delta_M = \Delta_C - \mu_D$, where Δ_C is the difference in the electron affinities of the Al MG and black-P BL and the black-P and the GL. For the black-P BL with the Al MG, $\Delta_M = 85$ meV, $\Delta_C = 225$ meV. The above condition of the proper band alignment is satisfied at $\mu_D = 140$ meV. This corresponds to

$\Sigma_D \simeq 1.6 \times 10^{12}\,\text{cm}^{-2}$. At elevated bias gate voltages V_G, the electron Fermi energy in the GC μ can somewhat exceed μ_D.

The main distinction of the device under consideration here and that studied previously is the method of the THz radiation input. We assume that the antenna system provides the signal voltages at the source and drain contacts shifted by the phase θ. According to the latter, the signal component GC potential $\delta\varphi_\omega$ with respect to the MG potential obeys the following conditions at the edges of the source ($x = -L$) and drain ($x = L$) contacts: $\delta\varphi_\omega|_{\pm L} = \delta V_\omega \exp(-i\omega t \pm i\theta/2)$. Here δV_ω and ω are the amplitude and the frequency of the electric signal provided by the antenna system under the effect of the impinging THz radiation, the axis x is directed in the GC plane from the source to the drain, and $2L$ is the spacing between these contacts (approximately equal to the GC length).

3. Equations of the Model

The signal voltages at the source and the drain excite the ac electric field δE_ω in the GC causing the absorption of the signal energy by the electrons. The absorbed power per unit of the GC area averaged over the radiation period (the averaging is denoted by the symbol $\langle\ldots\rangle$) is equal to the Joule power $Q_\omega = \text{Re}\sigma_\omega\langle|\delta E_\omega|^2\rangle$, where σ_ω is the GC ac Drude conductivity. The radiation absorption results in the variation, $\langle\delta T_\omega\rangle$, of the effective electron temperature in the GC. The latter leads to the variation of the thermionic currents from the GC via the BL into the MG [5,6]:

$$\langle\delta J_\omega\rangle^C = H j^{\text{max}} F \int_{-L_C}^{+L_C} dx \frac{\langle\delta T_\omega\rangle}{T_0}. \tag{5.1}$$

Here $F = (\Delta_M/T_0)\exp(-\Delta_M/T_0)$ is "the thermionic emission factor", $j^{\text{max}} = e\Sigma_D/\tau_\perp$ is the maximal value of the current density from the GC to the MG, $\tau_\perp$ is the electron try-to-escape time, and H is the GC width.

In the following, we consider the GC-FETs with $L_C \ll L$. In these GC-FETs, the plasmonic oscillations resonances in the whole gated GC are primarily attenuated by the electron scattering frequency ν in the GC encapsulated in the h-BN, which can be fairly small [9, 10], while the effect of a narrow black-AsP window section is weak.

For the axis x spatial distribution of the Joule power associated with the ac electric field produced by the plasmonic oscillations in the gated GC [13–15], one can obtain

$$\langle Q_\omega \rangle = \frac{\sigma_0}{2} \left(\frac{\delta V_\omega}{L} \right)^2 \left(\frac{\pi \nu}{2\Omega_P} \right)^2 \frac{\omega}{\sqrt{\nu^2 + \omega^2}}$$

$$\times \left| \left[\frac{\cos(\gamma_\omega x/L)\sin(\theta/2)}{\sin \gamma_\omega} - \frac{\sin(\gamma_\omega x/L)\cos(\theta/2)}{\cos \gamma_\omega} \right] \right|^2 .$$

$$(5.2)$$

Here $\gamma_\omega = \pi\sqrt{\omega(\omega + i\nu)}/2\Omega_P$ and $\Omega_P = (\pi e/\hbar L)\sqrt{\mu_D W/\kappa}$ is the plasmonic frequency with W and κ being the thickness and dielectric constant of the h-BN/b-P/h-BN BL, respectively. The right-hand side of Eq. (5.2) describes the resonant overshoots at $\omega \simeq \Omega_P$ and at higher plasmonic resonances.

The spatial distribution of $\langle \delta T_\omega \rangle$ is described by the one-dimensional electron heat transport equation, which for the device structure under consideration is presented in the following form [5, 6]:

$$-h\frac{d^2\langle \delta T_\omega \rangle}{dx^2} + \frac{\langle \delta T_\omega \rangle}{\tau_\varepsilon} = \langle Q_\omega \rangle. \qquad (5.3)$$

Here $h \simeq v_W^2/2\nu$ is the electron thermal conductivity in the GC per electron (this corresponds to the Wiedemann–Franz relation), $v_W \simeq 10^8\,\mathrm{cm/s}$ is the characteristic electron velocity in GCs, τ_ε is the electron energy relaxation time in the GC. Equation (5.3) describes the electron heat propagation along the GC and accounts for the electron heat transfer to the source and drain contacts. The latter can be significant due to high values of the electron heat conductivity [16, 17]. The second term in the left-hand side of Eq. (5.3) reflects the energy loss by the GC due to its transfer to the lattice [18–22]. The term describing the electron thermal energy leakage via the BL [5, 6, 23] is disregarded because of the small length, $2L_C$, of the window section [11]. Equation (5.3) is supplemented by the boundary conditions: $\langle \delta T_\omega \rangle|_{x=\pm L} = 0$.

Since the ac electric field along the GC is not a particularly strongly varying function of the coordinate x, one can replace the right-hand side of Eq. (5.3) by its spatially averaged value $\overline{\langle Q_\omega \rangle}$ (the Joule power averaged over the GC). Considering the signal frequencies close to the frequency of the fundamental plasmonic resonance and that the quality factor of this resonance $(2\Omega_P/\pi\nu)^2 \gg 1$, we obtain

$$\overline{Q_\omega} \simeq \sigma_0 \left(\frac{\delta V_\omega}{L}\right)^2 \left[\cos^2\left(\frac{\theta}{2}\right) + \left(\frac{\pi\nu}{4\Omega_P}\right)^2 \sin^2\left(\frac{\theta}{2}\right)\right]. \quad (5.4)$$

Using Eq. (5.4), we arrive at the approximate solution of Eq. (5.3) with the above boundary conditions in the following form:

$$\langle \delta T_\omega \rangle \simeq \overline{Q_\omega}\tau_\varepsilon \left[1 - \frac{\cosh(x/\mathcal{L})}{\cosh(L/\mathcal{L})}\right], \quad (5.5)$$

where $\mathcal{L} = \sqrt{h\tau_\varepsilon} = v_W\sqrt{\tau_\varepsilon/2\nu}$ is the electron cooling length. Equation (5.5) yields

$$\int_{-L_C}^{L_C} dx \langle \delta T_\omega \rangle \simeq 2\overline{Q_\omega}\tau_\varepsilon \left[L_C - \mathcal{L}\frac{\sinh(L_C/\mathcal{L})}{\cosh(L/\mathcal{L})}\right]$$

$$\simeq 2L_C\overline{Q_\omega}\tau_\varepsilon \left[1 - \frac{1}{\cosh(L/\mathcal{L})}\right]. \quad (5.6)$$

4. Characteristics

The GC-FET bolometric detector voltage responsivity at the fundamental plasmonic resonance is given by $R^V_{\omega=\Omega_P} = \langle \delta J_{\Omega_P}/S_{\Omega_P}\rangle r_L$, where S_{Ω_P} is the power of the normally impinging THz radiation received by the antennas and r_{Load} is the load resistance assumed to be equal to the GC-MG resistance. The quantity $(\delta V_\omega)^2 \propto S_\omega \cos^2 \psi$ (see, [24]), where ψ is the angle of the radiation incidence. In the case of the normal incidence ($\psi = 0$), using Eqs. (5.1) and (5.5) with Eq. (5.3), we arrive at the following formula:

$$R^V_{\omega=\Omega_P} \simeq R^V \left[\cos^2\left(\frac{\theta}{2}\right) + \left(\frac{\pi\nu}{4\Omega_P}\right)^2 \sin^2\left(\frac{\theta}{2}\right) \right]. \qquad (5.7)$$

Here

$$R^V = \frac{32}{137g}\left(\frac{\hbar}{eT_0}\right)\left(\frac{\Delta_M}{\mu}\right)\left(\frac{v_W^2 \tau_\varepsilon}{L^2\nu}\right)\left[1 - \frac{1}{\cosh(L/\mathcal{L})}\right]. \qquad (5.8)$$

It is worth noting that R^V is independent of $\tau_\perp$. This is because the GC- to MG THz radiation stimulated current is proportional to $\tau_\perp^{-1}$, while the GC-MG resistance (and, hence, the chosen load resistance) is proportional to $\tau_\perp$.

Figure 2 shows the phase dependence of the resonant voltage responsivity $R^V_{\omega=\Omega_P}$ calculated for the GC-FET bolometric detectors with the band parameters indicated above and characterized by different values of the electron collision frequency ν, i.e., different electron mobilities M. Additional parameters are listed in Table 1. One can see that the resonant voltage responsivity can be fairly high at certain values of the phase difference θ, especially in the

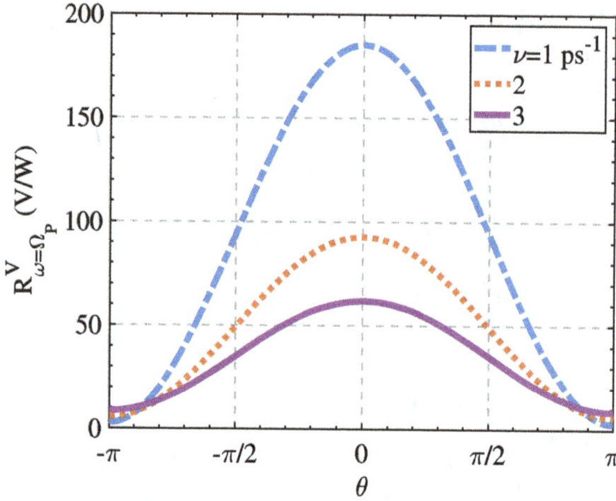

Fig. 2.   The GC-FET detector resonant responsivity $R^V_{\omega=\Omega_P}$ as function of phase difference θ at different values of collision frequencies ν.

Table 1.   GC-FET parameters.

GC length, $2L$	$1.5\,\mu$m
Gate layer thickness, W	$17.4\,$nm
Electron Fermi energy, μ	$140\,$meV
Plasmonic frequency, $\Omega_P/2\pi$	$1.0\,$THz
Lattice temperature, T_0	$25\,$meV
Energy relaxation time, τ_ε	$10\,$ps
Antenna gain, g	1.64

GC-FETs with relatively small collision frequencies ν (compare, for example, with [25, 26]). The electron mobility in the GC with the values of ν used in Fig. 2 can be estimated as $M \simeq (2.4 - 7.1) \times 10^4\,$cm^2/Vs, which is reasonable for GCs encapsulated in n-BN layers.

If the phase shift is associated with the inclined radiation incidence, its value is equal to $\theta = \omega\,2L_A \sin\psi/c = 4\pi\,L_A \sin\psi/\lambda_\omega$, where $2L_A$ is the distance between the antennas producing the voltage signals $\delta V_\omega \exp[-i(\omega t \mp \theta/2)]$ [not shown in Fig. 1(a)] at the source and drain, respectively, c is the speed of light in vacuum, and λ_ω is the THz radiation wavelength. Then we obtain

$$R^V_{\omega=\Omega_P} \simeq R^V \cos^2\psi \left[\cos^2(l_A \sin\psi) + \left(\frac{\pi\nu}{4\Omega_P}\right)^2 \sin^2(l_A \sin\psi)\right],$$

$$(5.9)$$

where $l_A = L_A\Omega_P/c$ is proportional to the ratio of the antenna spacing and the impinging THz radiation resonant wavelength and g is the antenna gain. The phase shift can be also associated with the circular polarization of the incident THz radiation [27].

Figure 3 demonstrates the angular dependence of the resonant responsivity $R_\omega = \Omega_P^V$ of the detectors with different spacing, L_A, between the antennas calculated for different values of collision frequency ν. As seen in Fig. 3, the angular width of the responsivity peaks is fairly small. This is confirmed by the values of the peaks full width of half maximum (FWHM) shown in Fig. 4.

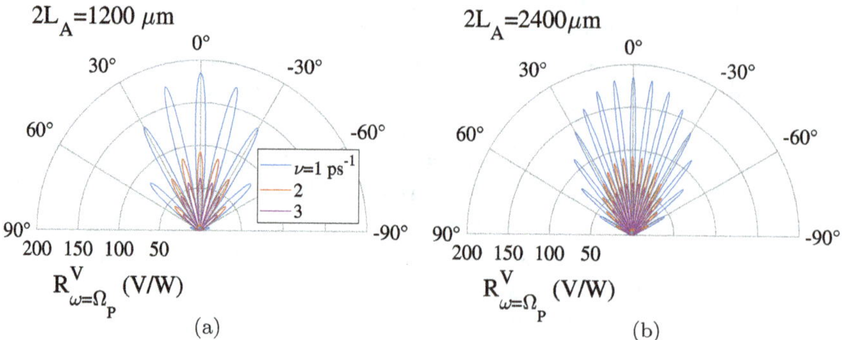

Fig. 3.   The GC-FET detector resonant responsivity $R^V_{\omega=\Omega_P}$ as a function of the angle of incidence ψ for different collision frequencies ν for different spacing between the antennas (a) $2L_A = 1200\,\mu$m and (b) $2L_A = 2400\,\mu$m.

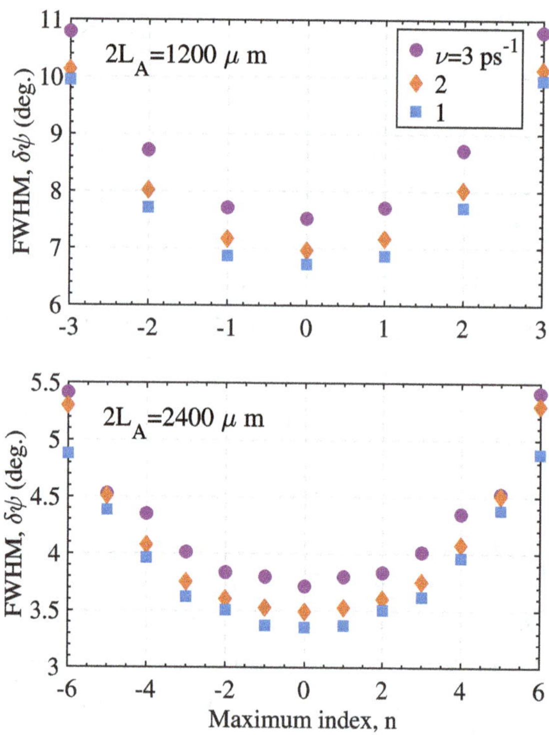

Fig. 4.   Angular FWHM of the different peaks of resonant responsivity $R^V_{\omega=\Omega_P}$ corresponding to Fig. 3: upper panel – $2L_A = 1200\,\mu$m and lower panel – $2L_A = 2400\,\mu$m.

Due to the voltage dependence of the plasmonic frequency Ω_P (associated with the variation of the Fermi energy), the plasmonic resonance can be voltage tuned. This might lead to the controllability of the responsivity phase and incidence angle.

Conclusion

In conclusion, we have proposed the THz GC-FET bolometric detectors with the h-BN/black-P/h-BN BLs and predicted that these detectors can exhibit fairly high room temperature responsivity with a pronounced phase and angular selectivity. Since the detector response strongly depends on the plasmonic resonance (i.e., the ratio ω/Ω_P), the latter is also very sensitive to the impinging radiation frequency.

Acknowledgments

The work at TU, UoA, and UB was supported by the Japan Society for Promotion of Science (KAKENHI Nos. 21H04546 and 20K20349), Japan and the RIEC Nation-Wide Collaborative Research Project No. R04/A10, Japan. The work at RPI was supported by AFOSR (contract number FA9550-19-1-0355).

ORCID

Victor Ryzhii https://orcid.org/0000-0002-5000-3109

Chao Tang https://orcid.org/0000-0002-6788-9575

Taiichi Otsuji https://orcid.org/0000-0002-0887-0479

Michael S. Shur https://orcid.org/0000-0003-0976-6232

Maxim Ryzhii https://orcid.org/0000-0001-5580-5971

Vladimir Mitin https://orcid.org/0000-0002-5579-5803

References

1. Xi Ling, H. Wang, S. Huang, F. Xia and M. S. Dresselhaus, "The renaissance of black phosphorus", *PNAS* **22** (2015) 4523–4530.
2. F. Xia, H. Wang and Y. Jia, "Rediscovering black phosphorous as an anisotropic layered material for optoelectronics and electronics", *Nat. Commun.* **5** (2014) 4458.
3. F. Liu, X. Zhang, P. Gong, T. Wang, K. Yao, S. Zhu and Y. Lu, "Potential outstanding physical properties of novel black arsenic phosphorus $As_{0.25}P_{0.75}/As_{0.75}P_{0.25}$ phases: A first-principles investigation", *RSC Adv.* **12** (2022) 3745–3754.
4. Y. Cai, G. Zhang and Y.-W. Zhang, "Layer-dependent band alignment and work function of few-layer phosphorene", *Sci. Rep.* **4** (2015) 6677.
5. V. Ryzhii, C. Tang, T. Otsuji, M. Ryzhii, V. Mitin and M. S. Shur, "Effect of electron thermal conductivity on resonant plasmonic detection in the metal/black-AsP/graphene FET terahertz hot-electron bolometers", *Phys. Rev. Appl.* **19** (2023) 064033.
6. V. Ryzhii, C. Tang, T. Otsuji, M. Ryzhii, V. Mitin and M. S. Shur, "Hot-electron resonant terahertz bolometric detection in the graphene/black-AsP field-effect transistors with a floating gate", *J. Appl. Phys.* **133** (2023) 174501.
7. Y. Liu *et al.*, "Mediated colossal magnetoresistance in graphene/black phosphorus heterostructures", *Nano Lett.* **18** (2018) 3377-3383.
8. A. S. Mayorov, R. V. Gorbachev, S. V. Morozov, L. Britnell, R. Jalil, L. A. Ponomarenko, P. Blake, K. S. Novoselov, K. Watanabe, T. Taniguchi and A. K. Geim, "Micrometer-scale ballistic transport in encapsulated graphene at room temperature", *Nano Lett.* **11** (2011) 2396–2399.
9. M. Yankowitz, Q. Ma, P. Jarillo-Herrero and B. LeRoy, "Van der Waals heterostructures combining graphene and hexagonal boron nitride", *Nat. Rev. Phys.* **1** (2019) 112–125.
10. M. Ryzhii, V. Ryzhii, M. S. Shur, V. Mitin, C. Tang and T. Otsuji, "Terahertz bolometric detectors based on graphene field-effect transistors with the composite h-BN/black-P/h-BN gate layers using plasmonic resonances", *J. Appl. Phys.* **134** (2023) 084501.
11. X. You, Ch. Fumeaux and W. Withayachumnankul, "Tutorial on broadband transmissive metasurfaces for wavefront and polarization control of terahertz waves", *J. Appl. Phys.* **131** (2022) 061101.
12. L. Banszerus, T. Sohier, A. Epping, F. Winkler, F. Libisch, F. Haupt, K. Watanabe, T. Taniguchi, K. Müller-Caspary, N. Marzari, F. Mauri, B. Beschoten and C. Stampfer, "Extraordinary high room-temperature carrier mobility in graphene-WSe2 heterostructures", arXiv:1909.09523v1.
13. V. Ryzhii, A. Satou and T. Otsuji, "Plasma waves in two-dimensional electron-hole system in gated graphene heterostructures", *J. Appl. Phys.* **101** 024509 (2007).

14. V. Ryzhii, T. Otsuji and M. S. Shur, "Graphene based plasma-wave devices for terahertz applications", *Appl. Phys. Lett.* **116** (2020) 140501.

15. A. V. Muraviev, S. L. Rumyantsev, G. Liu, A. A. Balandin, W. Knap and M. S. Shur, "Plasmonic and bolometric terahertz detection by graphene field-effect transistor", *Appl. Phys. Lett.* **103** (2013) 181114.

16. T. Y. Kim, C.-H. Park and N. Marzari, "The electronic thermal conductivity of graphene", *Nano Lett.* **16** (2016) 2439–2443.

17. Z. Tong, A. Pecchia, C. Yam, T. Dumitrică and T. Frauenheim "Ultrahigh electron thermal conductivity in T-Graphene, Biphenylene, and Net-Graphene", *Adv. Energy Mater.* **12** (2022) 2200657.

18. J. H Strait, H. Wang, S. Shivaraman, V. Shields, M. Spencer and F. Rana, "Very slow cooling dynamics of photoexcited carriers in graphene observed by optical-pump terahertz-probe spectroscopy", *Nano Lett.* **11** (2011) 4902–4906.

19. V. Ryzhii, M. Ryzhii, V. Mitin, A. Satou and T. Otsuji, "Effect of heating and cooling of photogenerated electron-hole plasma in optically pumped graphene on population inversion", *Jpn. J. Appl. Phys.* **50** (2011) 094001.

20. V. Ryzhii, T. Otsuji, M. Ryzhii, M. Ryzhii, N. Ryabova, S. O. Yurchenko, V. Mitin and M. S. Shur, "Graphene terahertz uncooled bolometers", *J. Phys. D: Appl. Phys.* **46** (2013) 065102.

21. D. Golla, A. Brasington, B. J. LeRoy and A. Sandhu, "Ultrafast relaxation of hot phonons in graphene-hBN heterostructures", *APL Mater.* **5** (2017) 056101.

22. K. Tamura, C. Tang, D. Ogiura, K. Suwa, H. Fukidome, Y. Takida, H. Minamide, T. Suemitsu, T. Otsuji and A. Satou, "Fast and sensitive terahertz detection in a current-driven epitaxial-graphene asymmetric dual-grating-gate FET structure", *APL Photonics* **7** (2022) 126101.

23. J. F. Rodriguez-Nieva, M. S. Dresselhaus and L. S. Levitov, "Thermionic emission and negative dI/dV in photoactive graphene heterostructures", *Nano Lett.* **15** (2015) 1451–1456.

24. R. E. Colin *Antenna and Radiowave Propagation*, McGraw-Hill, New York, 1985.

25. A. Rogalski, "Semiconductor detectors and focal plane arrays for far-infrared imaging", *Opto-Electron. Rev.* **21** (2013) 406–426.

26. A. Rogalski, M. Kopytko and P. Martyniuk, "Two-dimensional infrared and terahertz detectors: Outlook and status," *Appl. Phys. Rev.* **6** (2019) 021316.

27. I. V. Gorbenko, V. Yu. Kacharovskii and M. S. Shur, "Plasmonic helicity-driven detector of terahertz radiation," *Phys. Status Solidi RRL* **13** (2018) 1800464.

Chapter 6

Grating-Gate AlGaN/GaN Plasmonic Crystals for Terahertz Waves Manipulation[#]

M. Dub[*,†,§], P. Sai[†], A. Krajewska[†], D. B. But[†],
Yu Ivonyak[†], P. Prystawko[†], J. Kacperski[†], G. Cywiński[†],
S. Rumyantsev[†], W. Knap[†], M. Słowikowski[‡], and
M. Filipiak[‡]

[†]*Centera Laboratories, Institute of High Pressure Physics PAS,*
ul. Sokolowska 29/37, 01-142 Warsaw, Poland
[‡]*Centre for Advanced Materials and Technologies,*
CEZAMAT, Warsaw University of Technology,
ul. Poleczki 19, 02-822 Warsaw, Poland
[§]*maksimdub19f94@gmail.com*

The grating-gate plasmonic crystal system represents a compelling arena for investigating strong light-matter interactions and diverse plasmon resonances. This study reviews the recent discovery of two distinctive terahertz phases of AlGaN/GaN plasmonic crystals that arise from varying the modulation of a two-dimensional electron

---

[*]Corresponding author.
[#]This chapter appeared previously on the International Journal of High Speed Electronics and Systems. To cite this chapter, please cite the original article as the following: M. Dub, P. Sai, A. Krajewska, D. B. But, Y. Ivonyak, P. Prystawko, J. Kacperski, G. Cywiński, S. Rumyantsev, W. Knap, M. Słowikowski and M. Filipiak, *Int. J. High Speed Electron. Syst.*, **33**, 2440020 (2024), doi:10.1142/S0129156424400202.

density beneath the metallic gratings: the delocalized phase at weak modulation and the localized phase at strong modulation. Notably, we delve into an impact of the grating filling factor on the electrically driven transition between these phases. Our findings underscore the critical role of specific metal grating geometry parameters in facilitating this transition. Moreover, we explore the potential of utilizing graphene-based gratings as alternatives to metallic gratings. Through the integration of graphene, grown by Chemical Vapor Deposition method on copper foil and then transferred to the high electron mobility AlGaN/GaN heterostructures, we achieve an effective modulation of broadband absorption by free charge carriers within the 0.5–6 THz range via electrical biasing of the graphene electrode. However, while this approach successfully modulates absorption in a wide THz range, it does not elicit plasmon resonances within the graphene-based grating-gate plasmonic crystals. This intriguing observation poses a significant unresolved question warranting further theoretical and experimental exploration in subsequent studies.

Keywords: Plasmonic crystal; grating gate; AlGaN/GaN; graphene.

1. Introduction

Efficient manipulation of electromagnetic (EM) waves in the terahertz (THz) range plays a pivotal role in telecommunication technology [1, 2], optimization of wireless networks [3], broadening of medical applications [4], and fueling breakthroughs in scientific research. Various scientific approaches have emerged to exert control over THz radiation, leveraging light-matter interactions [5–7], artificial photonic bandgaps [8, 9], and metamaterials with negative refractive indices [10, 11] in photonic-crystal structures due to Bragg scattering from multiplied unit cells [12].

However, traditional photonic crystals based on bulk semiconductors, metals, and dielectrics exhibit limitations in tunability and reconfiguration of electromagnetic properties. The utilization of low-dimensional systems, such as two-dimensional electron gas (2DEG) within heterostructures, inverse channels, or graphene-like materials, presents a promising avenue. These materials offer a markedly more adaptable electromagnetic medium, amenable to tuning via external static electric fields. By introducing spatial periodicity to 2DEGs, a transformation into a plasmonic crystal occurs, effectively converting the 2DEG into a coherent plasmonic medium.

One feasible platform for realizing THz plasmonic systems is a field-effect transistor (FET) [13, 14] equipped with metal grating-gate electrodes [15, 16]. The grating-gate plasmonic crystal (GGPC) schematically illustrated in Fig. 1(a) emerges as a spatially periodic subwavelength resonant structure that enables a precise control of 2DEG concentration, while enhancing interaction between incoming EM radiation and 2D plasmons.

GGPCs have demonstrated enhanced performance as THz detectors [17–19], showcasing their potential utility in this domain. Ongoing experiments indicate a burgeoning role for GGPCs as prospective THz radiation emitters [20], amplifiers [21], and electrically controlled active THz modulators [22–24]. A recent discovery of distinct phases within these crystals [25] – a delocalized phase emerging at weak modulation of 2DEG density (Fig. 1(b)) and a localized phase manifesting at strong modulation (Fig. 1(c))– stimulated current investigation into the influence of metal grating

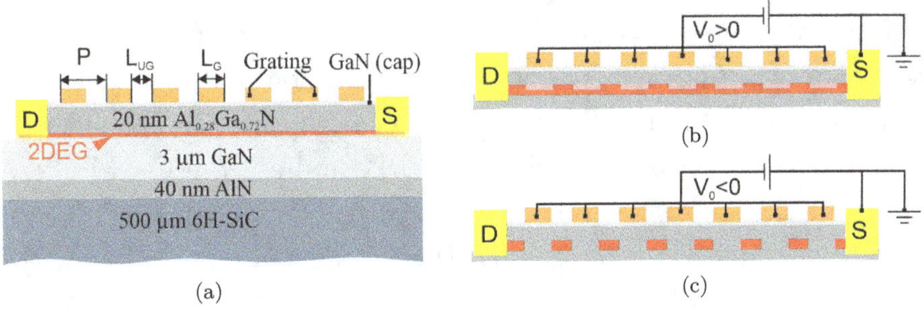

Fig. 1.   (a) Cross-section of AlGaN/GaN-based GGPC, where P is the grating period, L_G is the length of the gated region, L_{UG} is the length of the ungated region; (b) schematic representation of 2DEG concentration modulation by gate voltage applying in the regime of weak modulation (positive gate voltage swing: $V_0 > 0$); and (c) strong modulation (negative gate voltage swing: $V_0 < 0$). Source and drain electrodes are marked as S and D, respectively.

geometry parameters, particularly an impact of the grating filling factor on an electrical tuning of these phases and transitions between them.

Moreover, we experimentally explored the concept of replacing conventional metallic gratings with graphene-based alternatives and discussed the outcomes. Prior theoretical study [26] of graphene-based gratings suggests the potential for significantly enhancing THz wave amplification by several orders of magnitude, spurring our interest in their practical implementation and performance evaluation.

2. AlGaN/GaN-Based GGPCs: Fabrication and Characterization

Our sample fabrication involved utilizing AlGaN/GaN heterostructures boasting high 2DEG concentrations reaching up to 10^{13} cm^{-2} – an advantageous characteristic for THz plasmonics. This high electron concentration is particularly beneficial as the plasmon resonance frequency correlates proportionally with the square root of the electron concentration [27, 28], enabling a coverage across a broad spectrum of THz frequencies. The fabrication process employed Metalorganic Vapour Phase Epitaxy (MOVPE) within a closed coupled showerhead 3×2 inch Aixtron reactor (Aixtron, Herzogenrath, Germany). The epistructure comprised a sequence consisting of a 2 nm GaN cap, a 20 nm Al$_{0.28}$Ga$_{0.72}$N barrier, a 1 nm AlN layer (not depicted in Fig. 1(a)), and a 3 μm GaN buffer. This structure was grown directly on a 40 nm-thick AlN nucleation layer situated atop a 500 μm 6H-SiC substrate, as illustrated schematically in Fig. 1(a).

To facilitate an isolation of a device on the wafer, shallow 150 nm mesas were etched using Inductively Coupled Plasma-Reactive Ion Etching (ICP-RIE) equipment (Oxford Instruments, Bristol, UK). For the creation of ohmic drain and source contacts (identified as D for drain and S for source in Fig. 1), a thermal evaporation process involving a Ti/Al/Ni/Au metal stack of 150/1000/400/500 Å thickness was utilized. Subsequently, rapid thermal annealing (RTA) at 800°C in a nitrogen atmosphere for 60 s was conducted. This procedure resulted in establishing drain and source contacts with a contact resistance (R_c) of less than 1$\Omega\cdot$ mm, ensuring optimal electrical conductivity and efficient functionality of the device.

The final stage of processing involved an integration of the grating-gate electrode, a crucial step aimed at enlarging the active area of the gratings to the millimeter scale. This enlargement was imperative to ensure an efficient coupling between the relatively large THz wavelengths and the 2D plasmons. Simultaneously, maintaining the gate leakage current at negligible levels was pivotal for enabling the tuning of 2DEG concentration via the gate voltage (V_G) bias. All grating fingers were interconnected at a common base to facilitate simultaneous biasing of the entire structure. This specific configuration enabled for an exploration of 2D plasmon resonances within two distinct regimes, as depicted in Fig. 1(b): the regime characterized by a weak modulation of 2DEG concentration, corresponding to a positive gate voltage swing ($V_0 = V_G - V_{th}$, where V_{th} represents the threshold voltage of the channel depletion), and the regime demonstrating a strong modulation of 2DEG concentration (V_0 being negative).

Regarding the grating-gate electrodes, our investigation involved two distinct types of GGPCs. The initial type utilized conventional metallic gratings. For these structures, Schottky metallization was achieved through an electron beam lithography (EBL) for precise patterning, followed by a thermal evaporation of Ni/Au (150/350 Å) and subsequent metal lift-off. In contrast, the second type of device incorporated graphene-based gratings, a departure from the conventional Schottky metallization approach. These graphene-based gratings offered a novel alternative. Detailed geometrical parameters for all investigated samples are provided in Table 1, showcasing structures labeled S1–S6 representing regular

Table 1. Geometrical parameters of studied structures.

Structure parameters	S1	S2	S3	S4	S5	S6	S6G
Grating period, P (μm)	2	1.5	1	2.5	1.5	1	1
Gated region width, L_G (μm)	1.6	1.2	1	1.8	0.9	0.5	0.5
Ungated region width, L_{UG} (μm)	0.4	0.3	0.2	0.7	0.6	0.5	0.5
Grating filling factor, $r = L_G/P$	0.8	0.8	0.8	0.72	0.6	0.5	0.5
Active area, A (mm^2)				1.7×1.7			
Number of grating cells, N_{GC}	825	1100	1650	660	1100	1650	1650

metallic GGPCs, while the designation S6G denotes the graphene-based GGPC. Notably, all structures had a substantial active area of 1.7×1.7 mm^2 and different grating filling factors, r, from 0.5 to 0.8 (r = L_G/P, where L_G is the length of the gated region and P is the grating period). The optical microscope image of fabricated metallic grating-gates transistor shown in Fig. 2(a).

The fabrication of graphene-based gratings involved the use of commercially available single-layer graphene sourced from Graphenea, San Sebastián, Spain. This graphene was synthesized through the Chemical Vapor Deposition (CVD) method on thin copper foils [29]. To integrate graphene with AlGaN/GaN heterostructures, a rapid electrochemical delamination procedure [30] was employed for graphene transfer.

Initially, the graphene layers were transferred onto the entire active surface of the AlGaN/GaN sample. Subsequently, employing electron beam lithography (EBL), the graphene underwent patterning, followed by selective graphene etching in an oxygen plasma ambient environment within an ICP-RIE setup. Notably, a comparison between the segments featuring regular metallic and graphene-based gratings on the AlGaN surface is visualized in

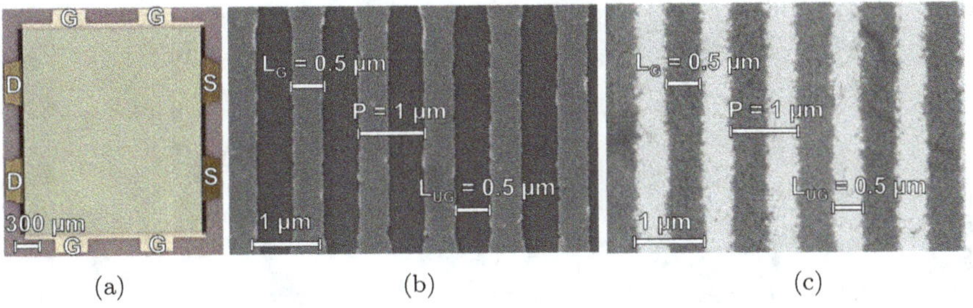

Fig. 2.   (a) Optical microscope image of the investigated grating-gate S6 sample. SEM images of metallic grating segment (S6 sample) (b), and graphene-based grating segment (S6G sample) (c). These samples are characterized by the same length of the gated region $L_G = 0.5$ μm, the length of the ungated region $L_{UG} = 0.5$ μm, and grating period $P = 1$ μm.

scanning electron microscope (SEM) images presented in Figs. 2(b) and 2(c) panels, respectively.

After fabrication, the samples were characterized by current-voltage measurements provided at 70 K. Figure 3 illustrates the channel current (blue lines) plotted against the gate voltage swing for both, a metallic grating S6 sample (panel a) and the graphene-based grating S6G sample (panel b). Notably, the red line depicted in each panel represents the 2DEG concentration (n_G) within the gated regions, acquired using an in-plane capacitor approximation to relate it to the gate voltage swing:

$$n_G = \frac{\varepsilon_{bar}\varepsilon_0 V_0}{ed}, \tag{6.1}$$

where d is the thickness of AlGaN barrier layer, ε_{bar} is the dielectric permittivity of barrier layer, and ε_0 is vacuum permittivity. Both metallic and graphene-based gratings exhibit comparable 2DEG concentrations within the gated regions, showcasing similar maximum values reaching 7×10^{12} cm^{-2} (as shown by red scattered lines in Fig. 3). Remarkably, both types of gratings are characterized by a minimized gate leakage current, enabling efficient tuning of n_G over a broad range via gate voltage application.

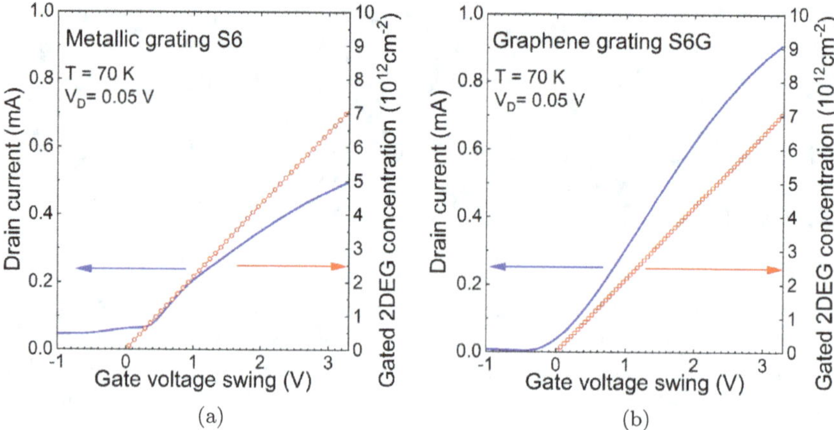

Fig. 3. Transfer current-voltage characteristics of the investigated metallic (a) and graphene-based (b) GGPCs measured at 70 K and drain-to-source voltage $V_D = 50$ mV. The red scattered lines show the gate voltage dependence of 2DEG concentration calculated using Eq. (1).

However, a notable distinction arises in the channel current between the two GGPC types. Specifically, the graphene-based GGPC displays nearly half the resistance at 70 K compared to its metallic counterpart, indicating a nearly twofold difference in the channel current (compare blue lines in panels a and b in Fig. 3). Taking into account that the ohmic contact resistance and the resistance of the ungated parts are the same for two types of GGPCs, this divergence can be attributed to the higher electron mobility observed in the AlGaN/GaN-based 2DEG channel beneath the graphene gate when compared to its metallic counterpart. The current-voltage measurements provided a possibility to evaluate the electron mobility under the gate, utilizing the following expression:

$$\mu_{2\text{DEG}} = \frac{L_G N_{GC} g_{m0}}{WC\left(V_D - IR_{acc}\right)}, \tag{6.2}$$

where N_{GC} is a number of grating cells (see Table 1, structures S6 and S6G), g_{m0} is the intrinsic transconductance calculated taking into account the access resistance R_{acc} consisting of the contact resistance and the resistance of ungated regions L_{UG}, V_D is drain voltage and I is channel current. This method provides $\mu_{2DEG} = 8800$ cm^2/V$\cdot$s for metallic and $\mu_{2DEG} = 16000$ cm^2/V$\cdot$s for graphene-based gratings extracted at $V_G = 0$. The higher electron mobility observed under the graphene gate may be attributed to suppressed scattering mechanisms. Further comprehensive investigations of different electron scattering mechanisms are required. However, such detailed analysis falls beyond the scope of this study.

3. Plasmon Resonances in GGPCs

GGPC includes plasmon oscillations within both gated and ungated periodically arranged sections. Notably, within gated plasma cavities, the plasmon frequency exhibits tunability by the gate voltage. Considering the proximity of the grating arrangement to the 2DEG channel and the gate voltage dependence of the electron concentration described by Eq. (1), the gate voltage dependence of 2D plasmon resonance frequency within the gated regions can be described by [27, 28]:

$$\omega_G = |k|\sqrt{\frac{eV_0}{m^*}}, \qquad (6.3)$$

where k is plasmon wave vector, e is the elementary charge, and m^* is the effective electron mass.

In the context of ungated cavities, a typical scenario involves a lack of the gate voltage tunability. Understanding the entire GGPC structure necessitates consideration of the interplay between plasma oscillations across different cavities. However, determining the behavior within this framework lacks a straightforward analytical solution. A more comprehensive approach entails employing the integral equation method [25] to solve the Maxwell set of equations, enabling a deeper understanding of the interactions between plasma oscillations in such intricate plasmonic crystal configurations.

To experimentally investigate the plasmon resonances in GaN-based GGPSs featuring different grating geometries (see Table 1), THz Fourier-transform spectrometry was employed. The transmission within the THz range was measured using a Fourier-transform infrared vacuum spectrometer (Vertex 80 v by Bruker, Billerica, Massachusetts, USA). The spectrometer setup comprised of a cryogenically cooled silicon bolometer with a solid-state silicon beam splitter, a mercury lamp as a source, and a cryostat utilizing a continuum flow of liquid Helium/Nitrogen (Optistat CF-V by Oxford Instrument, Abingdon, Great Britain). All measurements were done at 70 K. During measurements, a fast scanning mode was employed with a 1.5 mm aperture positioned directly after the samples, permitting electromagnetic radiation transmission solely through the grating-gate active region. The mirror moved at a frequency of 5 kHz, averaging interferograms over 100 scans within a double-scan interferogram regime. To mitigate Fabry-Pérot interferences stemming from the optically thick 500 μm SiC substrate, a spectrum resolution of 4 cm^{-1} (equivalent to 0.12 THz) was utilized.

The experimental results depicted in Fig. 4 showcase color counterplots illustrating the measured THz transmittance, delineating the transmission at specific gate voltage biases normalized to the signal observed without the sample. These plots are represented as a function of gate voltage swing for all the investigated metallic gratings. GGPC reveals two distinct THz phases – the recently discovered delocalized and localized phases [25].

The delocalized phase manifests itself during a weak modulation of 2DEG concentration (at positive values of V_0). Within this phase, 2D plasmons oscillate across both the gated and ungated sections of the GGPC, engaging in interactions. Notably, multiple orders of plasmon resonances were observed in all investigated metallic gratings. The fundamental resonance (indicated as (1) in Fig. 4) frequency depends on the grating period: the smaller the period, the higher the frequency for the fundamental resonance.

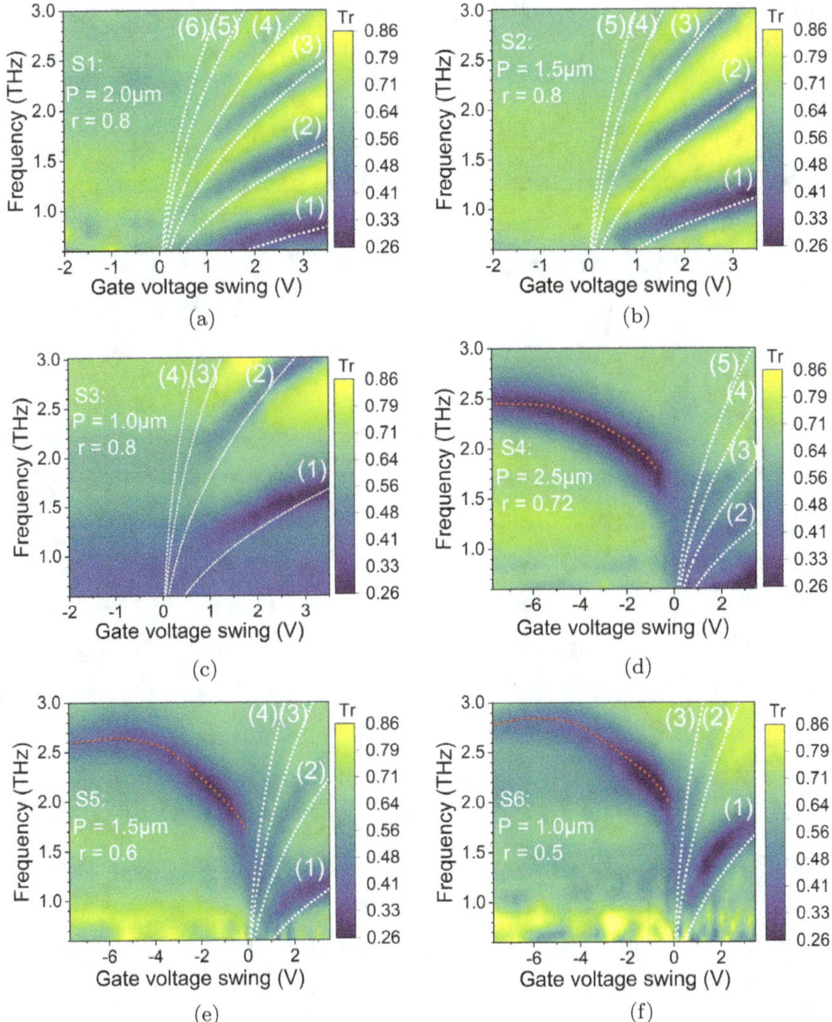

Fig. 4. Experimental contour plots of the transmittance (Tr) of the samples with metallic gratings of different geometry: (a) S1 sample; (b) S2 sample; (c) S3 sample; (d) S4 sample; (e) S5 sample; (f) S6 sample. White dotted lines are calculated using Eq. (3). Red dotted lines show the position of the localized GGPC phase extracted from experimental results.

Taking the plasmon wave vector controlled by the grating reciprocal vector ($k = 2\pi n/P$, $n = 1, 2, 3 \ldots$) and using Eq. (3), it is possible to evaluate the frequency of all plasmon resonances in the delocalized phase. In Fig. 4, these calculations are represented by the white dotted lines with the number denoting the respective order of each resonance. Notably, a better agreement was achieved for samples featuring the highest grating filling factor ($r = 0.8$; depicted in panels a, b, and c of Fig. 4). This observation suggests that deviations in the case of other samples may be attributed to the impact of ungated sections, which wield a more pronounced influence, particularly noticeable in samples characterized by smaller grating filling factors. Additionally, a higher grating filling factor allows the enhancement of plasmon resonances of higher orders, which is visible when comparing samples of the same grating period, but the different filling factors (samples S2 and S5; and samples S3 and S6).

Under conditions of strong modulation of 2DEG concentration (negative values of V_0), samples S4 (Fig. 4(d)), S5 (Fig. 4(e)), and S6 (Fig. 4(f)) undergo a transition toward the localized phase of GGPC. Within this phase, the gated regions of 2DEG are completely depleted, causing 2D plasmons to exclusively oscillate within the ungated portions of the GGPC. Remarkably, the frequency of plasmon resonances in this phase exhibits an unexpected dependency on the gate voltage. This observed blue shift in frequency, occurring with an increasingly negative V_0, can be explained by an additional depletion effect affecting the ungated cavities in the lateral direction – a phenomenon called the "edge gating effect" [31, 32].

Contrary to samples S4, S5, and S6 exhibiting a transition to the localized phase within the range of investigation (0.5–3 THz), the remaining samples – S1–S3 (depicted in Figs. 4(a)–4(c)) – do not demonstrate this transition within the same frequency range. The absence of this transition in S1–S3 could be attributed to the specific widths chosen for the gated and ungated regions. Notably, the frequency of resonances within the localized phase displays a direct correlation with the width of the ungated region – the narrower the widths, the higher the frequencies (Figs. 4(d) and 4(e)).

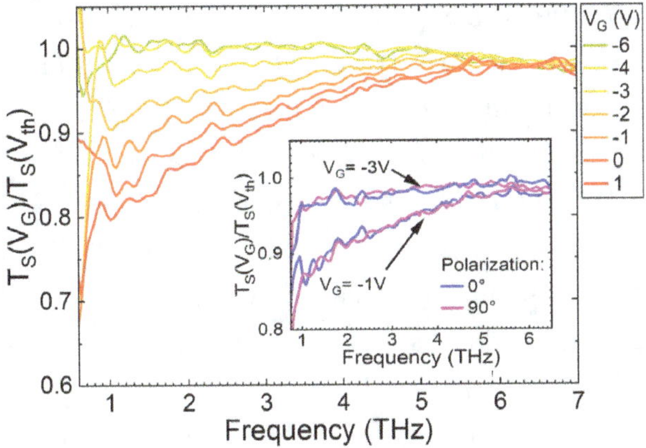

Fig. 5. THz transmission (T_S) spectra measured in graphene-based GGPC (sample S6G) and normalized to the transmission taken at $V_G = V_{\mathrm{th}}$. The insert illustrates the measured transmission at gate voltages $V_{G=-1}$ V and -3 V at different linear polarizations of incoming EM radiation corresponding to the grating fingers. All measurements are done at 70 K.

For samples S1–S3 (as detailed in Table 1), the smaller ungated regions ($\mathrm{L_{UG}}$) result in higher frequencies for localized resonances, extending beyond the examined frequency range. Consequently, these localized resonances do not interact with the experimentally measured plasmon resonances in the delocalized phase.

4. Graphene/AlGaN/GaN GGPC

In AlGaN/GaN GGPCs with graphene-based gratings (sample S6G), the THz transmission exhibits a strong dependence on gate voltage, as illustrated in Fig. 5. The figure showcases the THz transmission normalized against the threshold voltage of 2DEG depletion, revealing a pronounced broadband absorption within the 0.5–6 THz range, a behavior significantly influenced by gate voltages. This absorption phenomenon resembles Drude-like absorption attributed to free charge carriers, where higher 2DEG concentrations correspond to reduced transmittance. Furthermore, this Drude-like absorption persists at elevated temperatures, extending up to room temperature (not displayed in Fig. 5). Comparable observations of Drude absorption in Graphene were previously reported and studied in SiO_2 graphene devices [33, 34]. Our findings demonstrate the capability to modulate >15% of transmission within the 1–2 THz frequency range utilizing a single graphene layer integrated within a single AlGaN/GaN 2DEG layer. This remarkable modulation potential holds promise for developing broadband THz modulators based on graphene incorporated onto AlGaN/GaN structures.

However, plasmon resonances were not detected within the investigated THz range, remaining a significant unresolved aspect for prospective experimental and theoretical inquiries. Our primary hypothesis revolves around the inherent conductivity of CVD graphene transferred onto the AlGaN surface. Typically, such a graphene exhibits a sheet resistance of 1.5 kΩ/sq. [30], signifying considerably lower conductivity as compared to metallic grating conductivity. This difference in conductivity might contribute to the absence of observable plasmon resonances. Furthermore, we conducted polarization-dependent THz transmission measurements by integrating a linear polarizer into the spectrometer preceding the sample. As depicted in the insert in Fig. 5 showcasing the gates at $V_G = -1$ and -3 V, the results suggest that the CVD graphene grating does not polarize THz radiation as effectively as the metallic counterpart. This reduced polarization efficiency may consequently hinder the efficient coupling of THz waves and excitation of 2D plasmons.

5. Conclusion

In this study, the investigation into Grating-Gate Plasmonic Crystals (GGPCs) has unraveled distinct THz phases dependent on the modulation of 2DEG density beneath metallic gratings. Our investigation notably showcased the feasibility of observing a gate voltage-controlled transition between delocalized and localized phases within GGPCs.

This transition becomes evident when the frequency of 2D plasmon resonances in both the gated and ungated sections of the plasmonic crystal falls within the same spectral range. Notably, instances of gratings with a high filling factor reaching 0.8 did not exhibit this phase transition, underscoring the nuanced interplay between grating characteristics and phase transitions within GGPCs.

We studied regular metallic and novel graphene-based gratings for GGPCs. While metallic gratings presented with tunable plasmon resonances, the adoption of graphene-based gratings enabled for a control over a broadband of the free charge carrier absorption within the 0.5–6 THz range. However, the conspicuous absence of plasmon resonances in graphene-based GGPCs stands as a pertinent avenue for future theoretical and experimental exploration. This discrepancy is attributed to the diminished conductivity of CVD graphene and its reduced polarization efficiency compared to metallic counterparts.

Acknowledgments

This work was supported by the "International Research Agendas" program of the Foundation for Polish Science, co-financed by the European Union under the European Regional Development Fund for CENTERA Lab (No. MAB/2018/9) and co-funded by the European Union (ERC "TERAPLASM," Project No. 101053716). Views and opinions expressed are those of the author(s) only and do not necessarily reflect those of the European Union or the European Research Council. Neither the European Union nor the granting authority can be held responsible for them. P. S. is grateful for the partial support provided by the National Science Centre, Poland allocated on the basis of Grant No. 2019/35/N/ST7/00203.

ORCID

M. Dub https://orcid.org/0000-0002-6898-4638

P. Sai https://orcid.org/0000-0002-6143-9469

A. Krajewska https://orcid.org/0000-0001-6616-2932

D. B. But https://orcid.org/0000-0002-0735-4608

Yu Ivonyak https://orcid.org/0000-0002-7444-4056

P. Prystawko https://orcid.org/0000-0003-0762-5639

G. Cywiński https://orcid.org/0000-0003-3616-8756

S. Rumyantsev https://orcid.org/0000-0002-8308-4187

W. Knap https://orcid.org/0000-0003-4537-8712

M. Slowikowski https://orcid.org/0000-0001-7265-3199

M. Filipiak https://orcid.org/0000-0002-4978-4630

References

1. H.-J. Song and N. Lee, "Terahertz communications: Challenges in the next decade". *IEEE Transactions on Terahertz Science and Technology* **12**, 105 (2021).
2. A. Leitenstorfer, A. S. Moskalenko, T. Kampfrath, J. Kono, E. Castro-Camus, K. Peng, N. Qureshi, D. Turchinovich, K. Tanaka and A. G. Markelz, "The 2023 terahertz science and technology roadmap". *Journal of Physics D: Applied Physics* **56**, 223001 (2023).
3. T. Kürner, D. M. Mittleman and T. Nagatsuma, *THz communications: Paving the way towards wireless TBPS* (Springer, 2022).
4. L. Yu, L. Hao, T. Meiqiong, H. Jiaoqi, L. Wei, D. Jinying, C. Xueping, F. Weiling and Z. Yang, "The medical application of terahertz technology in non-invasive detection of cells and tissues: opportunities and challenges". *RSC Advances* **9**, 9354 (2019).
5. D. Nicoletti and A. Cavalleri, "Nonlinear light-matter interaction at terahertz frequencies". *Advances in Optics and Photonics* **8**, 401 (2016).
6. J. W. Han, P. Sai, D. B. But, E. Uykur, S. Winnerl, G. Kumar, M. L. Chin, R. L. Myers-Ward, M. T. Dejarld and K. M. Daniels, "Strong transient magnetic fields induced by THz-driven plasmons in graphene disks". *Nature Communications* **14**, 7493 (2023).
7. K.-J. Tielrooij, A. Principi, D. S. Reig, A. Block, S. Varghese, S. Schreyeck, K. Brunner, G. Karczewski, I. Ilyakov and O. Ponomaryov, "Milliwatt

terahertz harmonic generation from topological insulator metamaterials". *Light: Science & Applications* **11**, 315 (2022).

8. B. Reinhard, G. Torosyan and R. Beigang, "Band structure of terahertz metallic photonic crystals with high metal filling factor". *Applied Physics Letters* **92**, 201107 (2008).

9. F. Wu, M. Chen, Z. Chen and C. Yin, "Omnidirectional terahertz photonic band gap broaden effect in one-dimensional photonic crystal containing few-layer graphene". *Optics Communications* **490**, 126898 (2021).

10. G. Swift, A. Gallant, N. Kaliteevskaya, M. Kaliteevski, S. Brand, D. Dai, A. Baragwanath, I. Iorsh, R. Abram and J. Chamberlain, "Negative refraction and the spectral filtering of terahertz radiation by a photonic crystal prism". *Optics Letters* **36**, 1641 (2011).

11. A. Berrier, M. Mulot, M. Swillo, M. Qiu, L. Thylén, A. Talneau and S. Anand, "Negative refraction at infrared wavelengths in a two-dimensional photonic crystal". *Physical Review Letters* **93**, 073902 (2004).

12. G. Acuna, S. Heucke, F. Kuchler, H.-T. Chen, A. Taylor and R. Kersting, "Surface plasmons in terahertz metamaterials". *Optics Express* **16**, 18745 (2008).

13. M. Dyakonov and M. Shur, "Shallow water analogy for a ballistic field effect transistor: New mechanism of plasma wave generation by dc current". *Physical Review Letters* **71**, 2465 (1993).

14. W. Knap, F. Teppe, Y. Meziani, N. Dyakonova, J. Lusakowski, F. Boeuf, T. Skotnicki, D. Maude, S. Rumyantsev and M. Shur, "Plasma wave detection of sub-terahertz and terahertz radiation by silicon field-effect transistors". *Applied Physics Letters* **85**, 675 (2004).

15. V. Popov, M. Shur, G. Tsymbalov and D. Fateev, "Higher-order plasmon resonances in gan-based field-effect transistor arrays". *International Journal of High Speed Electronics and Systems* **17**, 557 (2007).

16. A. Muravjov, D. Veksler, V. Popov, O. Polischuk, N. Pala, X. Hu, R. Gaska, H. Saxena, R. Peale and M. Shur, "Temperature dependence of plasmonic terahertz absorption in grating-gate gallium-nitride transistor structures". *Applied Physics Letters* **96**, 042105 (2010).

17. D. But, N. Dyakonova, D. Coquillat, F. Teppe, W. Knap, T. Watanabe, Y. Tanimoto, S. Boubanga Tombet and T. Otsuji, "Thz double-grating gate transistor detectors in high magnetic fields". *Acta Physica Polonica A* **122**, 1080 (2012).

18. J. A. Delgado-Notario, W. Knap, V. Clericò, J. Salvador-Sánchez, J. Calvo-Gallego, T. Taniguchi, K. Watanabe, T. Otsuji, V. V. Popov and D. V. Fateev, "Enhanced terahertz detection of multigate graphene nanostructures". *Nanophotonics* **11**, 519–529 (2022).

19. K. Tamura, C. Tang, D. Ogiura, K. Suwa, H. Fukidome, Y. Takida, H. Minamide, T. Suemitsu, T. Otsuji and A. Satou, "Fast and sensitive terahertz detection with a current-driven epitaxial-graphene asymmetric dual-grating-gate field-effect transistor structure". *APL Photonics* **7**, 126101 (2022).

20. V. A. Shalygin, M. D. Moldavskaya, M. Y. Vinnichenko, K. V. Mare-myanin, A. A. Artemyev, V. Y. Panevin, L. E. Vorobjev, D. A. Firsov, V. V. Korotyeyev, A. V. Sakharov, E. E. Zavarin, D. S. Arteev, W. V. Lundin, A. F. Tsatsulnikov, S. Suihkonen and C. Kauppinen, "Selective terahertz emission due to electrically excited 2D plasmons in AlGaN/GaN heterostructure". *Journal of Applied Physics* **126**, 183104 (2019).
21. S. Boubanga-Tombet, W. Knap, D. Yadav, A. Satou, D. B. But, V. V. Popov, I. V. Gorbenko, V. Kachorovskii and T. Otsuji, "Room-temperature amplification of terahertz radiation by grating-gate graphene structures". *Physical Review X* **10**, 031004 (2020).
22. L. Wang, Y. Zhang, X. Guo, T. Chen, H. Liang, X. Hao, X. Hou, W. Kou, Y. Zhao and T. Zhou, *Nanomaterials* **9**, 965 (2019).
23. A. Di Gaspare, E. A. A. Pogna, L. Salemi, O. Balci, A. R. Cadore, S. M. Shinde, L. Li, C. di Franco, A. G. Davies and E. H. Linfield, "Tunable, grating-gated, graphene-on-polyimide terahertz modulators". *Advanced Functional Materials* **31**, 2008039 (2021).
24. N. Pala, D. Veksler, A. Muravjov, W. Stillman, R. Gaska and M. Shur, "Resonant detection and modulation of terahertz radiation by 2DEG plasmons in GaN grating-gate structures". *Resonant detection and modulation of terahertz radiation by 2DEG plasmons in GaN grating-gate structures* (IEEE, 2007), p. 570.
25. P. Sai, V. V. Korotyeyev, M. Dub, M. Słowikowski, M. Filipiak, D. B. But, Y. Ivonyak, M. Sakowicz, Y. M. Lyaschuk, S. M. Kukhtaruk, G. Cywiński and W. Knap, "Electrical tuning of terahertz plasmonic crystal phases". *Physical Review X* **13**, 041003 (2023).
26. S. A. Mikhailov, "Plasma instability and amplification of electromagnetic waves in low-dimensional electron systems". *Physical Review B* **58**, 1517 (1998).
27. A. Chaplik, *Surface Science Reports* **5**, 289 (1985).
28. V. V. Popov, "Plasmon excitation and plasmonic detection of terahertz radiation in the grating-gate field-effect-transistor structures". *Journal of Infrared, Millimeter, and Terahertz Waves* **32**, 1178 (2011).
29. C. Backes, A. M. Abdelkader, C. Alonso, A. Andrieux-Ledier, R. Arenal, J. Azpeitia, N. Balakrishnan, L. Banszerus, J. Barjon and R. Bartali, "Production and processing of graphene and related materials". *2D Materials* **7**, 022001 (2020).
30. M. Dub, P. Sai, A. Przewłoka, A. Krajewska, M. Sakowicz, P. Prystawko, J. Kacperski, I. Pasternak, G. Cywiński and D. But, "Graphene as a Schottky Barrier contact to AlGaN/GaN Heterostructures". *Materials* **13**, 4140 (2020).
31. G. Cywiński, I. Yahniuk, P. Kruszewski, M. Grabowski, K. Nowakowski-Szkudlarek, P. Prystawko, P. Sai, W. Knap, G. Simin and S. Rumyantsev, "Electrically controlled wire-channel GaN/AlGaN transistor for terahertz plasma applications". *Applied Physics Letters* **112**, 133502 (2018).

32. P. Sai, D. But, I. Yahniuk, M. Grabowski, M. Sakowicz, P. Kruszewski, P. Prystawko, A. Khachapuridze, K. Nowakowski-Szkudlarek and J. Przybytek, "AlGaN/GaN field effect transistor with two lateral Schottky barrier gates towards resonant detection in sub-mm range". *Semiconductor Science and Technology* **34**, 024002 (2019).
33. Z. Huang, Q. Han, C. Ji, J. Wang and Y. Jiang, "Broadband terahertz modulator based on graphene metamaterials". *AIP Advances* **8**, 035304 (2018).
34. K. Yu, J. Jeon, J. Kim, C. W. Oh, Y. Yoon, B. J. Kim, J. H. Cho and E. Choi, "Infrared study of carrier scattering mechanism in ion-gated graphene". *Applied Physics Letters* **114**, 083503 (2019).

Chapter 7

Quantifying Flat-Band Voltage in Si Metal-Oxide-Semiconductor Structures: An Evaluation via Terahertz Emission Spectroscopy (TES)[#]

Dongxun Yang and Masayoshi Tonouchi[*,†]

Institute of Laser Engineering, Osaka University,
2-6 Yamadaoka, Suita, Osaka 565-0871, Japan

[†]*tonouchi.masayoshi.ile@osaka-u.ac.jp*

Laser-induced Terahertz (THz) Emission Spectroscopy (TES) has demonstrated its potential utility in the realm of Metal-Oxide-Semiconductor (MOS) devices as an expedient and noncontact estimation methodology. Owing to its discerning response to the interface electric field, the amplitude of the THz emission peak in time-domain spectroscopy encapsulates rich information regarding MOS properties, notably the flat-band voltage. This paper concentrates on the precise quantitative estimation of the flat-band voltage within the Si MOS structure, elucidating the intricacies of the estimation process through the THz emission model.

Keywords: Si metal-oxide-semiconductor; TES; flat-band voltage.

---

[*]Corresponding author.
[#]This chapter appeared previously on the International Journal of High Speed Electronics and Systems. To cite this chapter, please cite the original article as the following: D. Yang and M. Tonouchi, *Int. J. High Speed Electron. Syst.*, **33**, 2440018 (2024), doi:10.1142/S0129156424400184.

     D. Yang & M. Tonouchi

1. Introduction

The Silicon Metal-Oxide-Semiconductor (Si MOS) structure consti-
tutes a fundamental element in contemporary semiconductor technology [1]. The expeditious and judicious estimation of pertinent
parameters within this structure, notably the interface potential
and flat-band voltages, is imperative [2, 3]. Terahertz Emission
Spectroscopy (TES) emerges as a potential noncontact methodology, facilitating the rapid and semi-quantitative characterization of
various semiconductor structures [4] including Si MOS structures,
with a specific emphasis on the expeditious estimation of critical
parameters [5]. In this paper, an innovative approach employs the
intrinsic correlation between the amplitude of terahertz (THz)
emissions and the externally applied DC bias to the Si MOS
structure, and the precise quantitative estimation of the flat-
band voltage within the Si MOS structure is elucidated. The
proposed method underscores its utility in the swift assessment
and discernment of semiconductor characteristics, presenting a
promising avenue for efficiency and efficacy in the semiconductor
industry.

Ultrafast charge transport leads to the THz radiation [6–12].
With an ultrafast laser pulse illuminating the semiconductor
to generate the photocarriers, the movement of the transient
photocarriers radiates the THz wave. According to the types of
ultrafast charge transport, the mechanism of the THz emission can
be mainly concluded as follows: (1) carrier drift current accelerated
by electric field [13, 14]; (2) diffusion current by density gradient
[15, 16]; (3) carrier diffusion by quasi-ballistic hot electron motion
[17], shown in Fig. 1. The drift current relies on the association
between the electric field, carrier mobility, and optical penetration
depth. During the fast transient period of approximately 200–300
fs, fs laser illumination induces a photocurrent that is directly
proportional to the laser intensity. The THz emission field can be

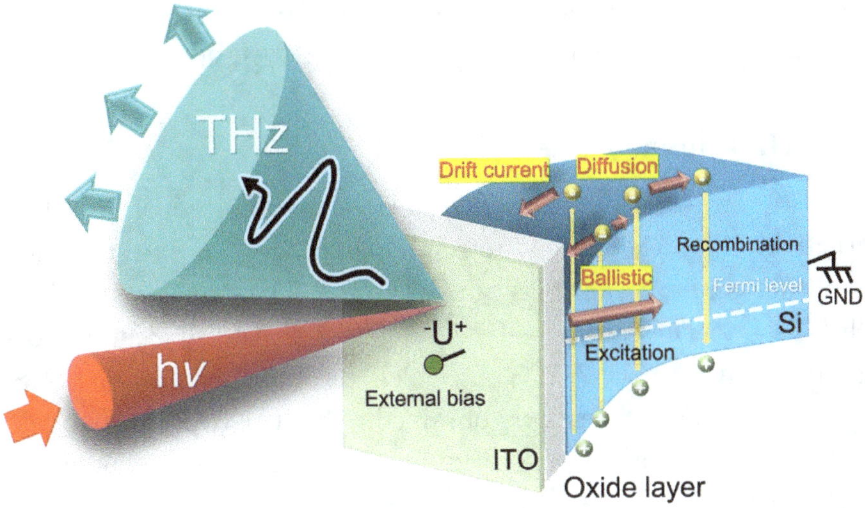

Fig. 1.   Mechanism of THz emission from semiconductors.

simply derived as

$$E_{\text{THz}} \propto \frac{dJ}{dt} \propto \frac{\Delta n \Delta v}{\Delta t} \propto \mu E_B I_p, \qquad (7.1)$$

where E_B is the built-in electric field, I_p is the intensity of the laser, and μ is the carrier mobility. The diffusion current includes the part by the density gradient or the photo-Dember effect and the part by the quasi-ballistic hot electron motion. When the energy of the hot photocarriers is significantly higher than the conduction bands, it leads to hot photocarrier injection. In such a scenario, the THz emission mechanism occurs due to quasi-ballistic transport of high-energy carriers and the THz emission field can be replaced by

$$E_{\text{THz}} \propto I_p \sqrt{\frac{E_p - E_g}{m^*}}, \qquad (7.2)$$

where E_p and E_g represent the photon energy and the bandgap of the semiconductor. m^* is the effective mass of the electrons.

Therefore, the THz emission due to the diffusion current is always constant when the laser power and samples have been decided.

2. THz Emission Theory Model

Here we provide the simplified THz emission model of MOS structure which elucidates the relationship between THz emission amplitude and external bias applied to Si MOS structures. Generally, the basic formula of THz emission from the semiconductors has been given in Eq. (7.1), in which the built-in field in the Si MOS structure can be simply obtained with depletion approximate assumption. Therefore, the THz emission field can be calculated from interface potential (V_s) as follows:

$$E_{\mathrm{THz}} \propto \frac{dJ}{dt} \propto \mu \frac{V_s}{w} I_p \propto \mu \sqrt{\frac{N_i V_s}{\varepsilon_0 \varepsilon_r}} I_p, \qquad (7.3)$$

where w is the thickness of depletion layer, N_i is the impurity density of Si, ε_0 and ε_r are the dielectric constants of the vacuum and Si.

In the ideal Si MOS structure, the external bias declines from the gate to the Si bulk where the electric potential is zero. Therefore, the external bias is equal to the sum of interface potential and potential drop across the oxide layer, expressed as follows:

$$V_g = V_s + V_{\mathrm{ox}}, \qquad (7.4)$$

where V_g is the external bias, V_s and V_{ox} are the interface potential and potential drop across the oxide layer. In addition, the capacitance of the oxide layer and the surface of the semiconductor are series connected; therefore, the relationship between V_{ox} and V_s is expressed as follows:

$$\frac{V_s}{V_{\mathrm{ox}}} = \frac{C_{\mathrm{ox}}}{C_s}, \qquad (7.5)$$

where $C_{\mathrm{ox}} = \varepsilon_0 \varepsilon_{\mathrm{ox}}/d$ is the capacitance of the oxide layer in a unit area. Based on these relations, V_g and V_s can be correlated as follows:

$$V_g = V_s \left(1 + \frac{C_s(V_s)}{C_{\mathrm{ox}}} \right). \qquad (7.6)$$

Equations (7.3) and (7.6) constitute the THz emission theory model of Si MOS, from which the relationship between the THz emission intensity and external bias voltages has been clarified [5]. In return, the parameters, including the interface potential and flat-band voltage of the Si MOS structure, can be estimated.

3. Experiment and Discussion

In this part, our focus revolves around the assessment of the Si MOS through the comparison of the experimental $A-V_g$ curve with the calculated $E_{\mathrm{THz}} - V_g$ curve, where the parameter A represents the amplitude of the first peak in the THz emission time-domain spectrum and E_{THz} represents the calculated THz emission field based on the THz emission theory model. The experiment was carried out in both n-type and p-type Si MOS structures under an 800-nm laser pulse illumination with external bias ranging from $-10\,\mathrm{V}$ to $10\,\mathrm{V}$. The measurement can also be achieved without ITO gate and external bias by using corona charging on the surface oxidized Si [18]. The samples were excited at $80\,\mathrm{MHz}$ using an 800-nm source at an incident angle of $45°\mathrm{C}$, and the THz emission was focused onto the GaAs THz sensor with variable delays. The average excitation power was set to $100\,\mathrm{mW}$. The details of the experimental results can be found in [13]. The calculated parameters of the laser, oxide layer, and Si are listed in Tables 1–3, respectively [13].

Based on the comparison in Fig. 2, significant disparities between calculated results and experimental observations are apparent along both the horizontal and vertical axes. These

Table 1.   Parameters of laser used in this study.

λ (nm)	r (mm)	P (mW)	I_p (W/m)^2
800	5	100	2546.55

Note: λ = laser wavelength; r = radius of laser beam; P = laser power; I_p = laser intensity.

Table 2.   Parameters of oxide layer.

	$\varepsilon_{\mathrm{ox}}$	d (nm)	ε_0 (F/m)	C_{ox} (mF/m^2)
p-type	3.9	120	8.854E$-$12	0.286
n-type	3.82	90	8.854E$-$12	0.374

Note: $\varepsilon_{\mathrm{ox}}$ = dielectric constant of oxide layer; d = thickness of oxide layer; ε_0 = vacuum dielectric constant; C_{ox} = capacitance per square meter of oxide layer.

Table 3.   Parameters of p- and n-type Si samples.

N_A (cm^{-3})	N_D (cm^{-3})	n_i (cm^{-3})	ε_r	$k(eV/K)$	E_g (eV)	$T(K)$	$V_{FB,p}$ (V)	$V_{FB,n}$ (V)
5.5E15	9.3E14	9.7E9	12	8.62E$-$5	1.12	300	-1.60	-0.96

Note: N_A = doping density of acceptors for p-type Si; N_D = doping density of donors for n-type Si; n_i = intrinsic carrier density; ε_r = dielectric constant of Si; k = Boltzmann constant; E_g = bandgap energy of Si; T = room temperature; $V_{\mathrm{FB},p}$ and $V_{\mathrm{FB},p}$ = flat-band voltages of the p- and n-type Si MOS samples, respectively.

disparities are primarily attributed to the flat-band voltage and the photo-Dember effect. Within our model, we posit an ideal condition wherein the flat-band voltage is assumed to be zero, and THz emission solely arises from the drift current. However, in practical Si MOS structures, the inherent work-function distinction between

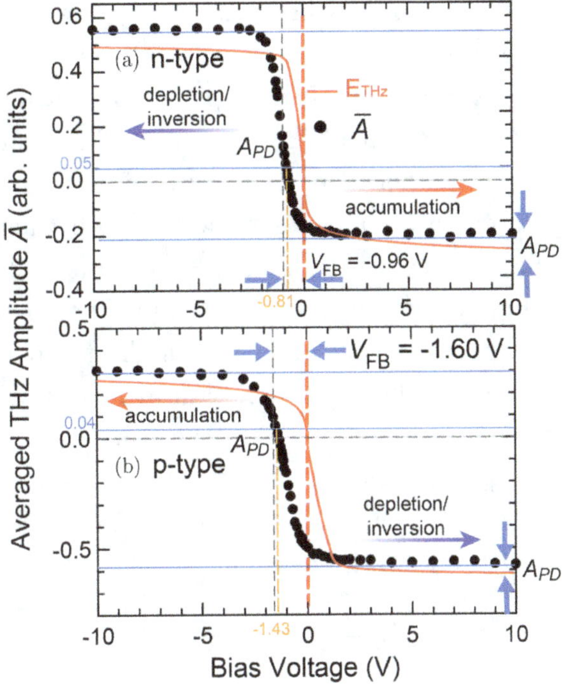

Fig. 2.   The calculation results of the $E_{\mathrm{THz}} - V_g$ curves in (a) n-type Si MOS and (b) p-type Si MOS and their relative $E_{\mathrm{THz}} - V_g$ curves. The original experimental data are taken from [13].

the gate material (ITO in our case) and the Si substrate induces initial interface band bending, thereby expediting the movement of laser-excited photocarriers and leading to THz wave radiation without external bias.

Another source of deviation along the vertical axis arises from THz emission resulting from the diffusion current due to the photo-Dember effect [15]. Laser illumination induces the generation of photocarriers in the semiconductor, and THz emission tied to the photo-Dember effect persists across various external bias voltages, including the flat band voltage condition. The intensity of this THz emission, contingent upon the bandgap of the semiconductor, remains constant. Hence, a rapid estimation of parameters can be facilely achieved by comparing the deviations between experimental and calculated results.

Here we introduce the detailed steps of the parameter estimation. First, standardization of E_{THz} and A. Since the values of calculated E_{THz} and experimental A are arbitrary units, we need to confirm the horizontal range is same for both calculation and experimental data, which means the absolute value of the difference between maximum value and minimum value should be standard and equivalent for both E_{THz} and A to make them comparable. Second, estimation of the THz emission from diffusion current (A_D). Roughly, we can directly obtain the value of A_D by putting the calculated E_{THz} and experimental A together into one chart and find out the difference between E_{THz} and A. We can also estimate the A_{PD} by using the following formula:

$$A_D = \frac{\sum_{i=1}^{n}\left(A_i - E_{\mathrm{THz},i}\right)}{n}. \tag{7.7}$$

The average of the difference between E_{THz} and A can be regarded as the value of the A_{PD}. When n becomes larger, the value of A_D becomes more accurate. Third, estimation of flat band voltage. At the flat band voltage, only the diffusion current takes charge of the THz emission. Therefore, we can easily obtain the flat band voltage from the $A - V_g$ curve when $A = A_D$,

$$V_{\mathrm{FB}} = V_g(A = A_D). \tag{7.8}$$

Table 4 delineates the flat-band voltage estimates derived from $A - V_g$ curve with the combination of the THz emission

Table 4. The estimation results of A_D and V_{FB} for both n- and p-type Si MOS.

	$A_{D,n}$ (a.u.)	$A_{D,p}$ (a.u.)	$V_{\mathrm{FB},n}$	$V_{\mathrm{FB},n}^{*}$	$V_{\mathrm{FB},p}$	$V_{\mathrm{FB},p}^{*}$
Value	0.05	0.04	-0.81 V	-0.96 V	-1.43 V	-1.60 V

Note: *The flat-band voltage estimated from C-V measurement.

theory model, alongside corresponding results obtained through C-V measurements. This approach enables a quantitative assessment of the flat-band voltage in Si MOS structures solely through TES, eliminating the necessity for supplementary methods like C-V measurements. This streamlined methodology enhances the practical utility of TES. Moreover, leveraging the calculated outcomes from the THz emission model, the quantitative association between E_{THz} and V_g offers the capacity to estimate additional parameters, including V_s and the various factors such as doping concentration of Si, dielectric constant and thickness of oxide layer, work-function of gate material [5, 19], etc. This extends the quantitative characterization capabilities of TES beyond the flat-band voltage, thereby providing a comprehensive insight into the intricate parameters governing Si MOS structures.

4. Conclusion

In conclusion, our investigation into Si MOS structures through TES has yielded valuable insights and quantitative estimations. By analyzing the experimental $A - V_g$ curve and comparing it with the calculated $E_{\mathrm{THz}} - V_g$ curve under ideal conditions, we successfully quantified the flat-band voltage for both n- and p-type Si MOS samples. Remarkably, our approach demonstrates the efficacy of TES as a standalone, noncontact method for flat-band voltage estimation, obviating the need for complementary techniques like C-V measurements. This methodology not only advances our understanding of fundamental MOS properties but also highlights the practical utility of TES in the semiconductor industry for rapid and accurate parameter estimation.

Acknowledgments

M.T. acknowledges support in part by JSPS KAKENHI Grant No. JP 23H00184, and JST, CREST Grant No. JPMJCR22O2,

Japan. The authors acknowledge T. Mochizuki *et al.* for providing experimental data and information.

ORCID

Dongxun Yang https://orcid.org/0000-0002-0546-1813

Masayoshi Tonouchi https://orcid.org/0000-0002-9284-3501

References

1. R. M. Jock, N. T. Jacobson, P. Harvey-Collard, A. M. Mounce, V. Srinivasa, D. R. Ward, J. Anderson, R. Manginell, J. R. Wendt, M. Rudolph, T. Pluym, J. K. Gamble, A. D. Baczewski, W. M. Witzel and M. S. Carroll, *Nat. Commun.* **9**(1), 1768 (2018).
2. T. Kobayashi, K. Tanaka, O. Maida and H. Kobayashi, *Appl. Phys. Lett.* **85**(14), 2806–2808 (2004).
3. G. Dubey, F. Rosei and G. P. Lopinski, *J. Appl. Phys.* **109**(10), 104904 (2011).
4. F. Murakami, A. Takeo, B. Mitchell, V. Dierolf, Y. Fujiwara and M. Tonouchi, *Commun. Mater.* **4**(1), 100 (2023).
5. D. Yang and M. Tonouchi, *J. Appl. Phys.* **130**(5), 055701 (2021).
6. V. Malevich, R. Adomavičius and A. Krotkus, *C. R. Phys.* **9**, 130–141 (2008).
7. A. Rustagi and C. J. Stanton, *Phys. Rev. B* **94**(19), 195207 (2016).
8. J. N. Heyman, N. Coates, A. Reinhardt and G. Strasser, *Appl. Phys. Lett.* **83**(26), 5476–5478 (2003).
9. D. Yang, A. Mannan, F. Murakami and M. Tonouchi, *Light Sci. Appl.* **11**(1), 334 (2022).
10. M. Tonouchi, *J. Appl. Phys.* **127**(24), 245703 (2020).
11. F. Murakami, A. Mannan, K. Serita, H. Murakami and M. Tonouchi, *J. Appl. Phys.* **131**, 185706 (2022).
12. Y. Huang, Z. Yao, C. He, L. Zhu, L. Zhang, J. Bai and X. Xu, *J. Phys.: Condens. Matter* **31**(15), 153001 (2019).
13. T. Mochizuki, A. Ito, J. Mitchell, H. Nakanishi, K. Tanahashi, I. Kawayama, M. Tonouchi, K. Shirasawa and H. Takato, *Appl. Phys. Lett.* **110**(16), 163502 (2017).
14. F. Song, X. Zu, Z. Zhang, T. Jia, C. Wang, S. Huang, Z. Liu, H. Xuan and J. Du, *J. Phys. Chem. Lett.* **13**(49), 11398–11404 (2022).
15. V. Apostolopoulos and M. E. Barnes, *J. Phys. D: Appl. Phys.* **47**(37), 374002 (2014).
16. G. Klatt, F. Hilser, W. Qiao, M. Beck, R. Gebs, A. Bartels, K. Huska, U. Lemmer, G. Bastian, M. B. Johnston, M. Fischer, J. Faist and T. Dekorsy, *Opt. Express* **18**(5), 4939–4947 (2010).

17. S. Preu, F. H. Renner, S. Malzer, G. H. Döhler, L. J. Wang, M. Hanson, A. C. Gossard, T. L. J. Wilkinson and E. R. Brown, *Appl. Phys. Lett.* **90**(21), 212115 (2007).
18. T. Mochizuki, A. Ito, H. Nakanishi, K. Tanahashi, I. Kawayama, M. Tonouchi, K. Shirasawa and H. Takato, *J. Appl. Phys.* **125**(15), 151615 (2019).
19. D. Yang, F. Murakami, S. Genchi, H. Tanaka and M. Tonouchi, *Appl. Phys. Lett.* **122**(4), 041601 (2023).

Chapter 8

Current-Driven Terahertz Oscillations in Diamond TeraFET[#]

Muhammad Mahmudul Hasan[*,†,§], **Nezih Pala**[†,¶], **and Michael Shur**[‡,‖]

†*Electrical and Computer Engineering,*
Florida International University,
10555 West Flagler Street Miami, Florida 33174, USA
‡*Department of Electrical Computer and Systems Engineering,*
110 8th Street Troy, New York 12180, USA
§*mhasa043@fiu.edu*
¶*npala@fiu.edu*
‖*shurm@rpi.edu*

We report on sub-terahertz plasmonic wave generation in the 2DEG channel of diamond TeraFET when biased by a DC current at the drain. Our numerical results demonstrated that p-diamond can support resonant oscillation of 300 GHz at room temperature, allowing it to function as a sub-THz emitter. We investigated the impact of different channel lengths, gate biases, drift velocities, and temperatures on the fundamental mode oscillation. The model incorporated the decay factors owing to electron scattering and electron fluid viscosity.

Keywords: THz source; sub-THz; p-Diamond.

---

*Corresponding author.

[#]This chapter appeared previously on the International Journal of High Speed Electronics and Systems. To cite this chapter, please cite the original article as the following: M. M. Hasan, N. Pala and M. Shur, *Int. J. High Speed Electron. Syst.*, **33**, 2440015 (2024), doi:10.1142/S0129156424400159.

1. Introduction

High-quality, compact, and room-temperature operational electronic terahertz sources and detectors will support next-generation THz applications in communication, imaging, security, sensing, and medicine [1]. Using plasma oscillations in the 2DEG channels of terahertz field-effect transistors (TeraFETs) is one of the key approaches for constructing a compact controllable THz radiation emitter [2]. Plasma waves are charge density oscillations in the 2D electron gas system in their channels. In recent years, different material systems like Si, III-N, graphene, and diamond have been investigated for TeraFET operation [3]. Among them, diamond was proposed as the stand-out material for plasmonic THz applications [4]. The large effective hole mass of diamond yields the longest momentum relaxation time among the other contemporary semiconductor materials. This allows diamond TeraFETs to fulfill the condition for resonant THz emission and detection even in the sub-THz frequencies at room temperature making it a great choice for 6G wireless communications applications. Resonant THz detection characteristics have been analyzed numerically for both p- and n-diamond TeraFET in the literature [5, 6]. Here, we report on the numerically investigated sub-THz emission from p- and n-diamond TeraFET exploiting the Dyakonov–Shur (DS) instability. We demonstrate room-temperature resonant plasma oscillations in the 2DEG channel around 300 GHz. We studied the emission characteristics of the p- and n-diamond TeraFETs for varying channel lengths, gate biases, and temperatures.

2. Simulation Methodology

We used a 1D model for the 2D hole (electron) gas system of the p-diamond (n-diamond) TeraFET'schannel (see Fig. 1). This nonlinear partial differential equation system of the hydrodynamic model was solved using COMSOL Multiphysics 6.0 [7,8].

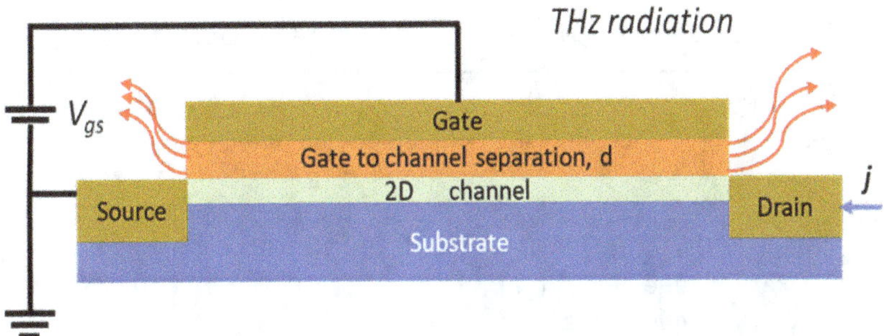

Fig. 1. Schematics of the simulated diamond TeraFET model with 2D channel [10].

$$\frac{\partial n_s}{\partial t} + \nabla \cdot (n_s V) = 0, \tag{8.1}$$

$$\frac{\partial V}{\partial t} + V\nabla V + \frac{e}{m}\nabla U + \frac{V}{\tau} - v\nabla^2 V = 0. \tag{8.2}$$

Equation (8.1) is the continuity equation where n_s is the charge carrier concentration, and Eq. (8.2) is the Euler equation [2]. The diffusion term, $-v\nabla^2 V$ is included in Eq. (8.2) where v is the viscosity of the carrier fluid. Also, the V/τ term accounts for the damping effect due to electron scattering with impurities and lattice vibrations, where τ represents the momentum relaxation time ($\mu = e\tau/m$; $\mu =$ mobility), V represents the carrier velocity in the channel, and m is the effective mass of the carrier. Also, in Eq. (8.2), U is the voltage difference between U_0 (gate bias above threshold/gate swing voltage) and U_{ch} (channel voltage). We assumed a constant gate swing voltage resulting in a constant carrier density, n_0, at the source side and a constant DC current density, meaning $j_0 = n_0 V_0$ at the drain side where V_0 is the drift velocity. The 2D charge density, n_s is calculated using a generalized formula following the unified charge control model (UCCM) [9].

$$n_s = \frac{C_g \eta U_{\text{th}}}{e} \ln\left(1 + \exp\left(\frac{U}{\eta U_{\text{th}}}\right)\right), \tag{8.3}$$

$$U_{\text{th}} = \frac{k_B T}{e}. \tag{8.4}$$

Table 1.   Material properties used in the simulation [11, 12].

Properties	p-diamond	n-diamond
Effective mass	0.74	0.36
Mobility (77 K) (m^2/Vs)	3.5	5
Mobility (300 K) (m^2/Vs)	0.53	0.73
Relaxation time (77 K) (ps)	14.73	10.24
Relaxation time (300 K) (ps)	2.23	1.49

Note: Above 300 K mobility values were extrapolated using $\mu \sim T^{-\frac{3}{2}}$ relation [11].

Here, C_g is the per unit area capacitance of the gate to channel barrier layer and η is an ideality factor. U_{th} represents the thermal voltage where k_B is the Boltzmann constant. The used material properties for p- and n-diamond in the simulation are listed in Table 1.

3.   Dyakonov–Shur Instability in Diamond TeraFET

We have simulated the instability of the diamond device for different channel lengths and gate swing voltages to tune the resonance plasma frequency. Gate to channel separation (d) was 4 nm. We observed the carrier oscillation at a position in the channel near the drain contact. The p-diamond model showed decaying carrier oscillation (see Fig. 2(a)) at low drift velocity. But with a higher drift velocity, the system can withstand continuous plasma wave oscillation in the channel as shown in Fig. 2(b). Here, the channel length, gate swing voltage, charge density, and temperature were 80 nm, 0.1 V, 7.92×10^{15} m^{-2}, and 300 K, respectively. At low drift velocity, the DS instability increment factor cannot overcome the damping from the electron scattering and the viscosity of the medium. But with high drift velocity, the increment factor crosses the damping barrier, and the channel can sustain the continuous oscillation of the charge density [10].

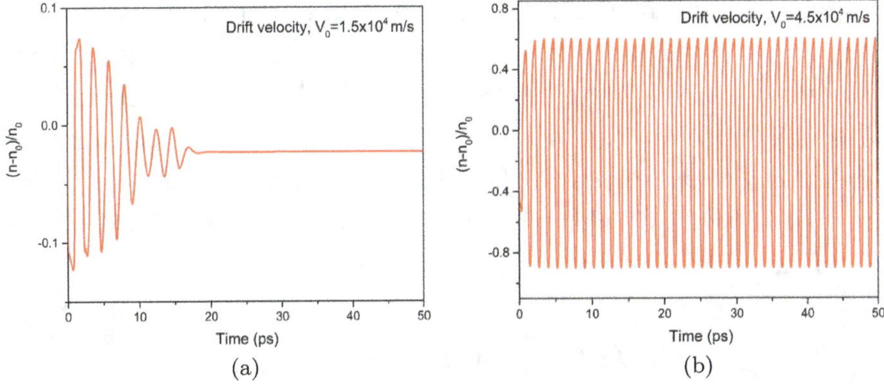

Fig. 2.   Charge density oscillation in p-diamond TeraFET (a) drift velocity, $V_0 = 1.5 \times 10^4\,\text{m/s}\,(1.9\,\text{mA/cm}^2)$ (b) drift velocity, $V_0 = 4.5 \times 10^4\,\text{m/s}$ $(5.7\,\text{mA/cm}^2)$; (channel length, $L = 80\,\text{nm}$, gate swing voltage, $U_0 = 0.1\,\text{V}$, Temperature, $T = 300\,\text{K}$).

The damping due to scattering and viscosity is calculated as $1/2\tau$ and $\pi^2 v/4L^2$. On the other hand, the increment of the plasma wave amplitude highly depends on the Mach number $(M = V_0/s)$ which is the ratio between the drift velocity (V_0) and the plasma wave velocity (s). The instability occurs when the drift velocity is high enough to make the net increment factor positive. The net increment is defined by subtracting the scattering and the viscosity losses from the initial increment. The plasma wave velocity, the increment, and the net increment can be calculated from Eqs. (8.5)–(8.7), respectively [2].

$$s = \sqrt{\left(\frac{eU_0}{m}\right)}, \tag{8.5}$$

$$\text{Increment}, \omega'' = \frac{s}{2L}(1 - M^2)\ln\left|\frac{1+M}{1-M}\right|, \tag{8.6}$$

$$\text{Net increment}, \omega'' = \frac{s}{2L}(1 - M^2)\ln\left|\frac{1+M}{1-M}\right|$$

$$- \frac{1}{2\tau} - \frac{\pi^2 v}{4L^2}. \tag{8.7}$$

For, channel length 80 nm, gate swing voltage 0.1 V, and viscosity $1.44\,\mathrm{cm^2/Vs^{12}}$, the threshold for the drift velocity to trigger the instability is $2.28 \times 10^4\,\mathrm{m/s}$ ($2.8\,\mathrm{mA/cm^2}$). As the velocity ($1.5 \times 10^4\,\mathrm{m/s}$; $1.9\,\mathrm{mA/cm^2}$) in Fig. 2(a) was below the threshold, the oscillation decayed to zero. However, a continuous plasma wave occurred in Fig. 2(b) when the velocity ($4.5 \times 10^4\,\mathrm{m/s}$; $5.7\,\mathrm{mA/cm^2}$) was higher than the threshold.

4. Resonance THz Oscillation

After recording the time domain signal, we transformed the signal into the frequency domain using the Fourier transform to get the oscillation frequency spectrum. To get the sustained plasma wave we maintained the drift velocity above the threshold level and below the plasma velocity to facilitate instability. We simulated both p- and n-diamond TeraFET models for different channel lengths, gate swing voltages, and temperatures to see the effect on the resonance frequency. The resonance frequency can be deduced from Eq. 8.8 [13].

$$f_0 = \frac{s^2 - V_0^2}{4Ls}.\tag{8.8}$$

4.1. *Effect of varying channel lengths and gate swing voltages*

The resonance frequency can be tuned using different channel lengths according to Eq. (8.8). For, p-diamond the simulation results showed resonant oscillation even up to 150 nm. The resonance frequency of p-diamond TeraFET moved in the sub-THz region from 800 GHz to 200 GHz for channel lengths 40 nm to 150 nm as shown in Fig. 3(a). We kept the drift velocity fixed at $4.5 \times 10^4\,\mathrm{m/s}$. The gate swing voltage was fixed at 0.1 V generating a charge density of $7.92 \times 10^{15}\,\mathrm{m^{-2}}$ at room temperature (300 K). p-diamond has a very large hole effective mass and long momentum relaxation time which makes it easier to achieve high quality factor even with 150 nm channel length to generate resonant plasma wave. The reduced amplitude for the longer channel lengths can be ascribed to the inverse proportional relation of the channel length with the increment factor. Results also showed that resonance oscillations are possible from 200 GHz to 300 GHz band for 100 nm to 150 nm p-diamond channel at 300 K.

The fundamental oscillation mode also depends on the plasma wave velocity through the gate swing voltage of the device. When we swept the gate swing voltage from 0.1 V to 1 V in the p-diamond TeraFET the resonance frequency increased from 250 GHz to almost 1 THz as shown in Fig. 3(b). Here, we used 120 nm channel length and room temperature conditions. The drift velocity was the same (4.5×10^4 m/s) as in the previous case. The value of current density changed from 1.9 mA/cm^2 to 56.7 mA/cm^2 for the increment of hole concentration from 7.29×10^{15} m^{-2} to 7.88×10^{16} m^{-2}. Higher gate swing voltage increases the plasma wave velocity and hence increases the fundamental mode frequency. We also analyzed the resonance plasma wave oscillation in the n-diamond TeraFET. Sweeping from shorter to longer channel lengths showed a similar trend (like p-diamond) in n-diamond as

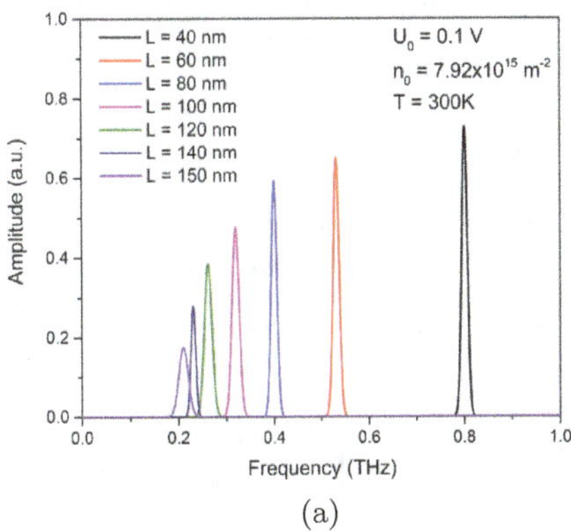
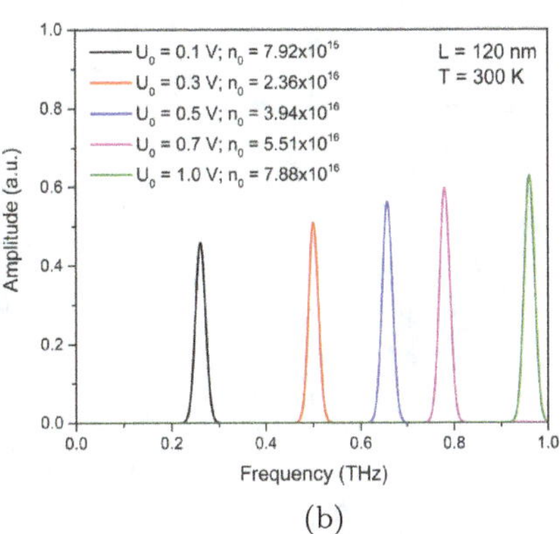

(a)   (b)

Fig. 3. Spectrum profile of p-diamond TeraFET for (a) varying channel lengths (L) ($U_0 = 0.1$ V, $n_0 = 7.92 \times 10^{15}$ m^{-2}, $T = 300$ K), and (b) different gate swing voltages (U$_0$) ($L = 120$ nm, T = $300\,K$).

well as shown in Fig. 4. However, using the same biasing conditions with the same channel length, n-diamond showed higher resonance frequency than p-diamond.

This happens as n-diamond has a lighter electron-effective mass than p-diamond. Also, n-diamond did not sustain the resonance plasma wave for the gate lengths longer than 130 nm, whereas p-diamond did so for the gate length of 150 nm. The momentum relaxation time of n-diamond is lower than p-diamond which made it harder for n-diamond to achieve high quality factor in the lower sub-THz band. The drift velocity was 5×10^4 m/s for 40 nm to 100 nm in n-diamond but it was increased to 6×10^4 m/s and 7×10^4 m/s for 120 nm and 130 nm to compensate for the reduced increment factor. It is also noticeable that the required drift velocities for generating DS instability in both p- and n-diamond TeraFET are well below their respective saturation velocities [14].

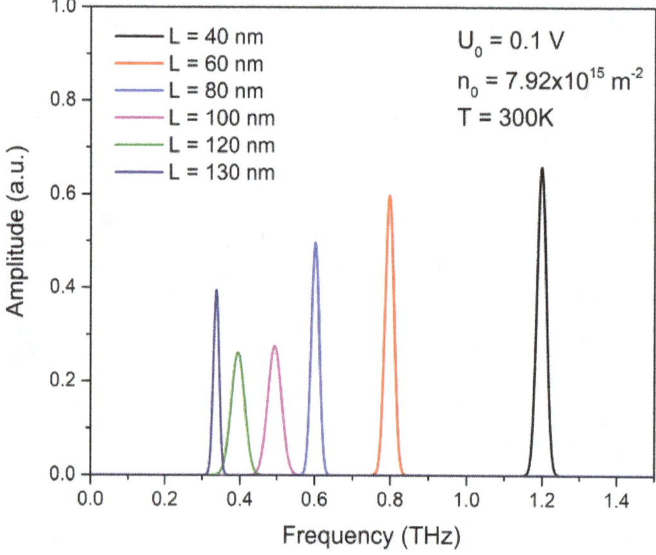

Fig. 4.   Spectrum profile of n-diamond TeraFET for varying channel lengths (L) ($U_0 = 0.1$ V, $n_0 = 7.92 \times 10^{15}$ m^{-2}, $T = 300$K).

4.2. *Temperature effect on resonance frequency*

We analyzed the effect of different temperature conditions on the resonance frequency for p- and n-diamond. The values of mobility and momentum relaxation time of p- and n-diamond at different temperatures were extrapolated using the $\mu \sim T^{-3/2}$ relation. At cryogenic (77 K) and room temperature (300 K), both p- and n-diamond oscillate at 250 GHz (see Fig. 5(a)) and 400 GHz (see Fig. 5(b)), respectively. Here, the channel length was 120 nm, and the gate swing voltage was 0.1 V. The reduced amplitude at room temperature (300 K) is due to the decreased mobility at high temperatures which shortens the momentum relaxation time. Also, n-diamond required higher drift velocities than p-diamond to achieve resonance at cryogenic and room temperature. Figure 5(a) shows that p-diamond can have resonance oscillation even at very high temperatures (500 K and 700 K) at higher frequencies. The resonance moves to higher frequencies as elevated gate swing (0.3 V and 1.0 V) voltage is required to achieve a large quality factor at high temperatures. The resonance was achieved at 700 K using the maximum saturation velocity of p-diamond [14]. A similar trend was noticed in n-diamond as well (see Fig. 5(b)). However,

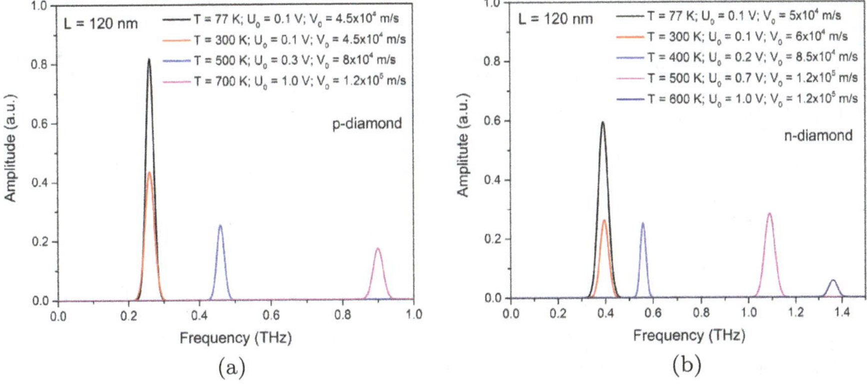

Fig. 5.   Resonance frequency at different temperatures with channel length 120 nm in (a) p-diamond, and (b) n-diamond TeraFET.

     M. M. Hasan, N. Pala & M. Shur

n-diamond could only reach $400\,\text{K}$ before crossing its saturation velocity $(8.5 \times 10^5\,\text{m/s})$ [14].

5. Conclusion

We numerically demonstrated room temperature plasma wave generation from p-diamond TeraFET around $300\,\text{GHz}$. Both p- and n-diamond showed room temperature resonance oscillation at the sub-THz band while p-diamond can sustain these oscillations for longer channel length $(150\,\text{nm})$ devices. Our analysis also predicts resonance oscillations in p-diamond at elevated temperatures $(700\,\text{K})$ at higher frequencies up to $0.9\,\text{THz}$. The resonance frequency depends on the channel length and carrier frequency, and hence, could be adjusted by gate bias. The results show that p-diamond is a promising material for 6G communications applications at $300\,\text{GHz}$ and beyond.

Acknowledgment

This work was supported in part by NSF under Award #: 1950788.

ORCID

Muhammad Mahmudul Hasan https://orcid.org/0000-0001-7955-9357

Nezih Pala https://orcid.org/0000-0001-6136-9811

Michael Shur https://orcid.org/0000-0003-0976-6232

References

1. T. Otsuji and M. Shur, "Terahertz plasmonics: Good results and great expectations," *IEEE Microwave Magazine,* vol. 15, no. 7, pp. 43–50, 2014.
2. M. Dyakonov and M. Shur, "Shallow water analogy for a ballistic field effect transistor: New mechanism of plasma wave generation by dc current," *Physical Review Letters,* vol. 71, no. 15, p. 2465, 1993.
3. M. Shur, "Si, SiGe, InP, III-N, and p-diamond FETs and HBTs for sub-terahertz and terahertz applications," in *Terahertz, RF, Millimeter, and*

Submillimeter-Wave Technology and Applications XIII, 2020, vol. 11279: International Society for Optics and Photonics, p. 1127903.

4. M. Shur, S. Rudin, G. Rupper and T. Ivanov, "p-Diamond as candidate for plasmonic terahertz and far infrared applications," *Applied Physics Letters,* vol. 113, no. 25, p. 253502, 2018.

5. Y. Zhang and M. S. Shur, "P-diamond Plasmonic TeraFET Detector," in *2020 45th International Conference on Infrared, Millimeter, and Terahertz Waves (IRMMW-THz)*, 2020: IEEE, pp. 1–2.

6. M. M. Hasan, Y. Zhang, N. Pala and M. Shur, "Diamond-based plasmonic terahertz devices," in *2023 IEEE 16th Dallas Circuits and Systems Conference (DCAS)*, 2023: IEEE, pp. 1–4.

7. S. Rudin, G. Rupper, A. Gutin and M. Shur, "Theory and measurement of plasmonic terahertz detector response to large signals," *Journal of Applied Physics,* vol. 115, no. 6, p. 064503, 2014.

8. "COMSOL Multiphysics 6.0," I. COMSOL, Ed., ed.

9. Y. H. Byun, K. Lee and M. Shur, "Unified charge control model and subthreshold current in heterostructure field-effect transistors," *IEEE Electron Device Letters,* vol. 11, no. 1, pp. 50–53, 1990.

10. M. M. Hasan, N. Pala and M. Shur, "Sub-Terahertz Plasma Wave Generation by Dyakonov-Shur Instability in p-Diamond TeraFET," in *2023 IEEE Photonics Conference (IPC)*, 2023: IEEE, pp. 1–2.

11. I. Akimoto, Y. Handa, K. Fukai and N. Naka, "High carrier mobility in ultrapure diamond measured by time-resolved cyclotron resonance," *Applied Physics Letters,* vol. 105, no. 3, p. 032102, 2014.

12. Y. Zhang and M. S. Shur, "p-Diamond, Si, GaN, and InGaAs TeraFETs," *IEEE Transactions on Electron Devices,* vol. 67, no. 11, pp. 4858–4865, 2020.

13. W. Knap *et al.*, "Terahertz emission by plasma waves in 60 nm gate high electron mobility transistors," *Applied Physics Letters,* vol. 84, no. 13, pp. 2331–2333, 2004.

14. M. Pomorski *et al.*, "Characterisation of single crystal CVD diamond particle detectors for hadron physics experiments," *Physica Status Solidi (A),* vol. 202, no. 11, pp. 2199–2205, 2005.

Chapter 9

Compact Spice Models for Terafets[#]

Xueqing Liu[†,¶], Trond Ytterdal[‡,‖] and Michael Shur[§,]**

[†]*Department of Electrical, Computer and Systems Engineering,*
Rensselaer Polytechnic Institute, 110 8th Street, Troy, NY 12180, USA
[‡]*Department of Electronic Systems,*
Norwegian University of Science and Technology,
7491 Trondheim, Norway
[§]*Department of Electrical, Computer and Systems Engineering,*
Rensselaer Polytechnic Institute, 110 8th Street, Troy, NY 12180, USA
[¶]*liux31@rpi.edu*
[‖]*trond.ytterdal@ntnu.no*
[**]*shurm@rpi.edu*

Field effect transistors (FETs) in plasmonic regimes of operation could
detect terahertz (THz) radiation and operate as THz interferometers,
spectrometers, frequency-to-digital converters and THz modulators
and sources. We report on the development of compact models
for Si MOS (Metal-Oxide-semiconductor) and heterostructure-based
plasmonic FETs (or TeraFETs) suitable for circuit design in the THz
range and based on the multi-segment unified charge control model.
This model accounts for the electron inertia effect (by incorporating
segmented Drude inductances), for the ballistic field effect mobility,
which is proportional to the channel length, for parasitic resistances
and capacitances and for the leakage current. It is validated by

---

[#]This chapter appeared previously on the International Journal of High Speed
Electronics and Systems. To cite this chapter, please cite the original article
as the following: X. Liu, T. Ytterdal and M. Shur, *Int. J. High Speed Electron.*
Syst., **33**, 2450004 (2024), doi:10.1142/S0129156424500046.

comparison with experimental data and TCAD simulation results. The model can be used for simulation and optimization of sub-THz and THz detectors. Our simulations use up to 200 segments in the device channel. The results are also in good qualitative agreement with the hydrodynamic simulations. Applications of our model could dramatically reduce astronomical design costs of nanoscale VLSI reaching US\$1.5 billion for the 3 nm technological node.

Keywords: THz; FETs; SPICE; compact model.

1. Introduction

The drive toward Beyond 5G (6G communication) technologies will rely on high-performance nanoscale transistors operating in the terahertz range (0.1–10 THz). Accurate compact models are critical for the design of nanoscale FET circuits, especially since the design costs are rapidly increasing with the development of advanced technology nodes reaching US\$1.5 billion for the 3 nm technological node [1]. The TCAD simulation [2], hydrodynamic modeling [3] and the recent FET SPICE models [4–8] revealed that the compact model in the THz range must use a distributed multi-segment channel model and include not only the distributed nonlinear channel capacitance and nonlinear distributed channel resistance but also the distributed Drude inductance accounting for the electron inertia. Such an approach allows us to accurately account for the quasi-ballistic transport at short dimensions and/or at high frequencies. We now present such a model for InGaAs HFETs (Heterostructure FETs), GaN-based HFETs and Si MOSFETs. The model is validated by comparing it with the hydrodynamic theory for THz detectors and the experimental data. The model validation also reveals the significance of electron inertia as the new physics becomes dominant in the submillimeter range.

2. THz SPICE Model and Validation

In a short channel TeraFET, incident THz radiation coupling to the contact edges could induce a THz AC voltage between the gate and source and gate and drain contacts. The THz AC voltage could modulate both the carrier density and the carrier drift velocity and thus excite plasma waves (electron density oscillations) in the channel [9, 10]. For non-symmetrical boundary conditions (i.e., short circuit at the source-gate and open circuit at the gate-drain) the rectification of the excited plasma waves leads to a DC response voltage at the drain, which could be either resonant or nonresonant depending on the THz signal frequency and the FET channel length. The analytical THz response was first derived with the hydrodynamic equations for the two-dimensional electron fluid in the channel of FETs [10]. It was also improved to include the extrinsic resistances and parasitic capacitances and account for the subthreshold regime and the ballistic field effect mobility, which is proportional to the channel length [7, 8].

Figure 1(a) shows the equivalent circuit of the one-segment THz SPICE model. The intrinsic FET nonlinear capacitances C_{gs} and C_{gd} of the model are calculated using the equations of the unified charge control model [11]. The SPICE model accounts for the extrinsic components including the parasitic capacitances C_{par_gs} and C_{par_gd} and the series resistances R_s and R_d. The model also includes the Drude inductance $L_{drude} = \tau R_{ch}$, where $\tau = m\mu/q$ is the electron momentum relaxation time, m is the electron effective mass, μ is the electron mobility and R_{ch} is the channel resistance [12].

L_{drude} accounts for the electron inertia effect critical for plasmonic resonant detection. The one-segment model could be applied at low THz frequencies for short channel devices yielding a good agreement with the experimental results [13]. However, at high THz frequencies, it is necessary to use the multi-segment model to achieve accurate results [6]. Figure 1(b) shows the equivalent circuit of the multi-segment SPICE model, where the intrinsic FET, internal nonlinear capacitances and the Drude inductance are split into N segments. The minimum number of segments could be estimated as $N \geq 3L/L_o$, where L is the channel length and

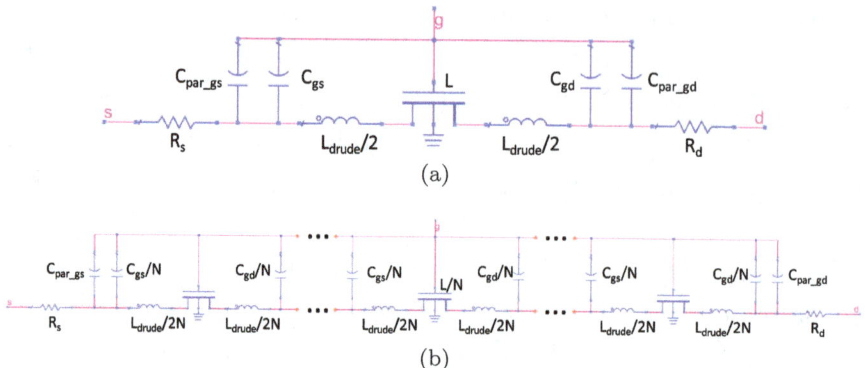

Fig. 1. Equivalent circuit for the TeraFET (a) one-segment model and (b) multi-segment model [4–6].

L_o is the characteristic decay length of the THz ac voltage away from the source, indicating the plasma wave penetration. Above the threshold, L_o is given by $L_o = (\mu V_{gt}/(2\pi f))^{1/2}$, where f is the THz radiation frequency and V_{gt} is the gate voltage swing [14].

The channel segmentation is significant in accounting for the strongly non-uniform electron density oscillations in the channel at high THz frequencies. Figure 2 shows the minimum number of segments needed for the multi-segment model with $L = 130\,\text{nm}$

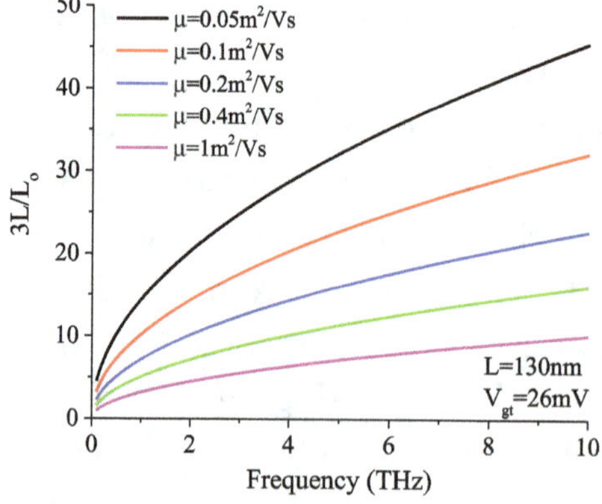

Fig. 2. Minimum number of segments needed for the multi-segment model with $L = 130\,\text{nm}$, $V_{gt} = 26\,\text{mV}$ for different mobilities.

and $V_{gt} = 26\,\mathrm{mV}$ for different mobilities. As seen, more segments are required at higher frequencies for devices with lower mobilities.

The static characteristics should be the same for the one-segment and the multi-segment model. This is achievable by tuning the SPICE parameters of the multi-segment model. Figure 3 shows the comparison of the simulated I-V characteristics for the

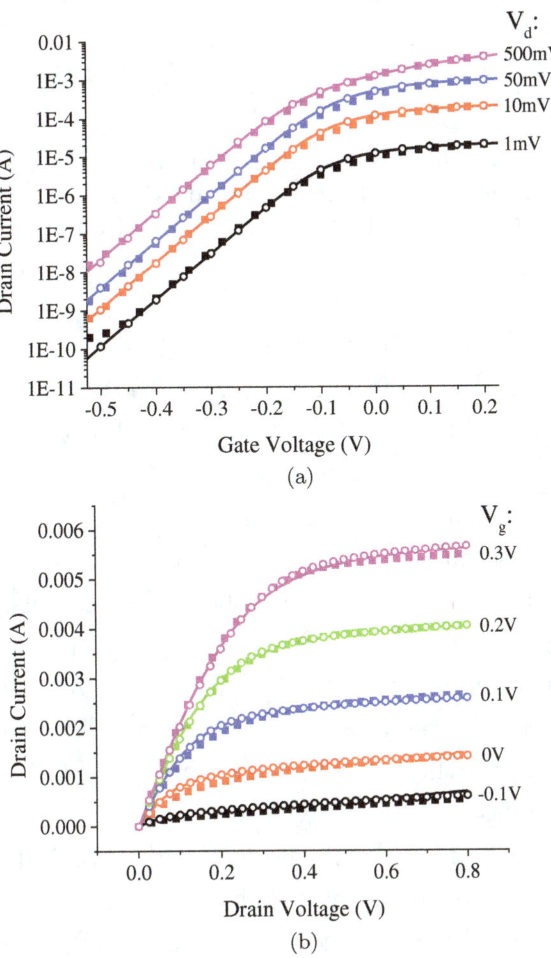

Fig. 3.   Comparison of the simulated I-Vs for the one-segment model (lines) and the multi-segment model (circles) with the measured I-Vs (squares) for AlGaAs/InGaAs p-HEMTs.

     X. Liu et al.

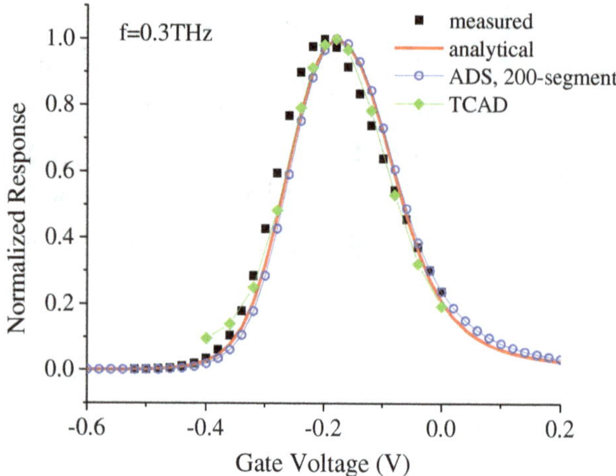

Fig. 4.   Comparison of the THz response as a function of the gate bias between the measurement, the analytical theory, the 200-segment model in ADS and the numerical model in TCAD [2] for the AlGaAs/InGaAs HFET.

one-segment and the multi-segment models with the measured I-Vs using AlGaAs/InGaAs p-HEMTs with 130 nm gate length and 18 μm gate width. The SPICE parameters for the one-segment and the multi-segment models were adjusted to achieve a good agreement with the measured results.

Using the determined parameters, the multi-segment SPICE model could be used to simulate the THz response. Figure 4 shows the dependence of the THz response on the gate bias for the 200-segment model from the harmonic balance simulation in ADS, compared with the analytical and measured results and our TCAD model [2] for the AlGaAs/InGaAs HFET. Good agreement validates the multi-segment SPICE model.

The multi-segment SPICE model is also validated for THz detector applications in other material systems. Figure 5 shows the comparison of the measured and simulated THz response and drain current versus the gate bias using the 200-segment model for AlGaN/GaN HFET and Si MOSFET. The good agreement demonstrates the applicability of the multi-segment SPICE model

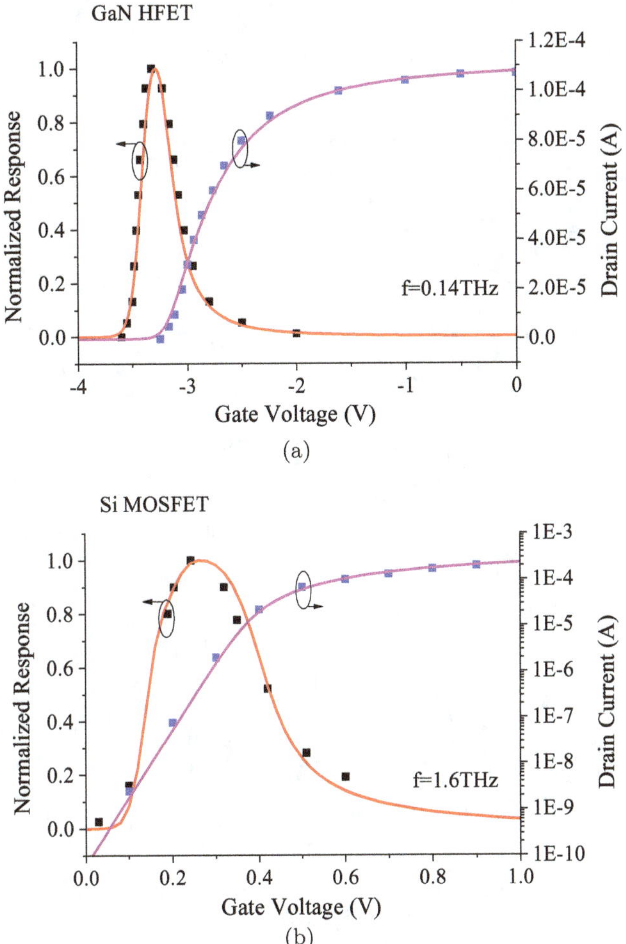

Fig. 5. Comparison of the THz response and the drain current between the 200-segment model (lines) and the measurement (squares) for (a) AlGaN/GaN HFET [15] and (b) Si MOSFET [16, 17].

for the THz detectors in different material systems for a wide range of the applied biases and THz frequencies.

The advantage of the multi-segment model is its validity at high THz frequencies. Figure 6 shows the comparison of the THz response as a function of the THz radiation frequency between the analytical theory and the SPICE models with one and 200 segments

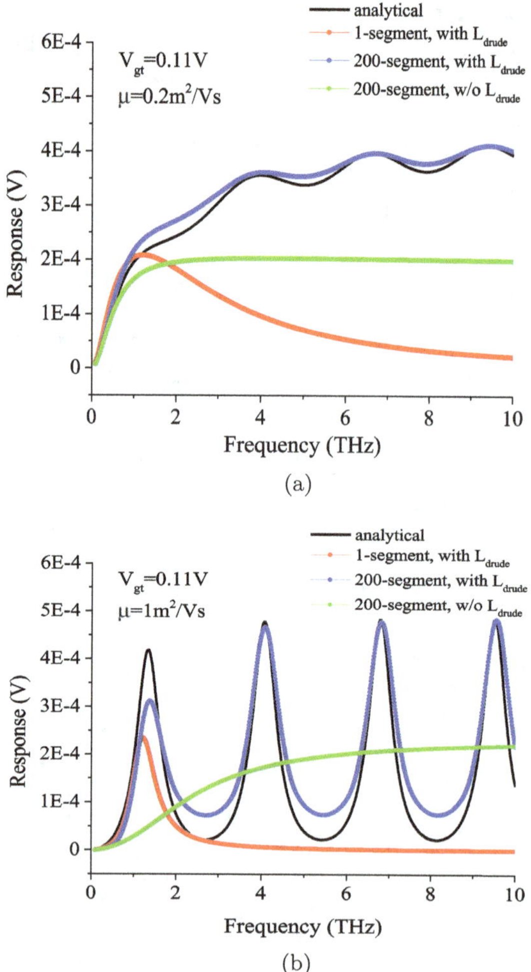

Fig. 6. THz response predicted by the analytical theory and the SPICE models for the AlGaAs/InGaAs HFET with (a) low mobility and (b) high mobility.

for the AlGaAs/InGaAs HFET with different mobilities. In this comparison, for the ease of exploring the effects of the channel segmentation and the Drude inductance, the models do not include the extrinsic components. The results show that at higher frequencies ($> 1\,THz$), the one-segment model deviates from the analytical theory and is no longer valid, while the multi-segment model has good agreement with the analytical theory in the wide THz frequency range. For devices with high mobilities, which could be achieved at cryogenic temperatures, the multi-segment model shows the resonances at the same resonant frequencies predicted by the analytical theory. Also, the results demonstrate that L_{drude} is critical for the THz SPICE model, especially at high frequencies.

3. Conclusion

The compact SPICE model described in this paper accounts for the electron inertia effect and includes the distributed channel resistances, capacitances and Drude inductances. Using up to 200 segments, the model was validated from the good agreement with the theory for THz detection using FETs and the experimental data for InGaAs HFETs, GaN HFETs and Si MOSFETs. The validation demonstrates that the model could be well applied for nanoscale FETs in different material systems in a wide range of applied biases and THz frequencies. Both the electron inertia and the channel segmentation are significant in making an accurate compact model for the design, simulation and characterization of THz devices for Beyond 5G technologies.

ORCID

Xueqing Liu https://orcid.org/0000-0002-4192-1557
Trond Ytterdal https://orcid.org/0000-0002-2109-833X
Michael Shur https://orcid.org/0000-0003-0976-6232

References

1. M. LaPedus, "5 nm Vs. 3 nm." *Semiconductor Engineering*, Jun. 2019. semiengineering.com/5 nm-vs-3 nm/
2. X. Liu and M. Shur, "TCAD model for TeraFET detectors operating in a large dynamic range," *IEEE Trans. Terahertz Sci. Technol.*, 10(1), 15–20, Nov. 2019.

3. H. Marinchio, G. Sabatini, C. Palermo, J. Pousset, J. Torres, L. Chusseau, L. Varani, P. Shiktorov, E. Starikov and V. Gružinskis, "Hydrodynamic modeling of optically excited terahertz plasma oscillations in nanometric field effect transistors," *Appl. Phys. Lett.*, 94(19), pp. 192109, May 2009.

4. A. Gutin, A. Muraviev, M. Shur, V. Kachorovskii and T. Ytterdal, "Large signal analytical and SPICE model of THz plasmonic FET," *in Lester Eastman Conf. on High Performance Devices (LEC), Singapore*, Singapore, 2012, pp. 1–5.

5. M. Shur, "Terahertz compact SPICE model." *in 2016 MIXDES-23rd Int. Conf. Mixed Design of Integrated Circuits and Systems*, 2016, pp. 27–31.

6. X. Liu, T. Ytterdal, V. Yu. Kachorovskii and M. S. Shur, "Compact terahertz SPICE/ADS model," *IEEE Trans. Electron Devices*, 66(6), pp. 2496–2501, Apr. 2019.

7. X. Liu, K. Dovidenko, J. Park, T. Ytterdal and M. S. Shur, "Compact Terahertz SPICE model: Effects of drude inductance and leakage," *IEEE Trans. Electron Devices*, 65(12), pp. 5350–5356, Dec. 2018.

8. X. Liu, T. Ytterdal and M. Shur, "TeraFET optimization for 0.1 THz to 10 THz operation," *URSI Radio Sci. Lett.*, 2, pp. 1–5, 2020.

9. M. Dyakonov and M. S. Shur, "Shallow water analogy for a ballistic field effect transistor: New mechanism of plasma wave generation by DC current," *Phys. Rev. Lett.*, 71(15), pp. 2465–2468, Oct. 1993.

10. M. I. Dyakonov and M. S. Shur, "Detection, mixing and frequency multiplication of terahertz radiation by two-dimensional electronic fluid," *IEEE Trans. Electron Devices*, 43(3), pp. 380–387, Mar. 1996.

11. K. Lee, M. Shur, T. Fjeldly and T. Ytterdal, Semiconductor device modeling for VLSI. *Princeton, NJ: Prentice Hall*, 1993.

12. M. Shur, "Plasma wave terahertz electronics," *Electron. Lett.*, 46(26), pp. 18–21, Dec. 2010.

13. A. Gutin, S. Nahar, M. Hella and M. Shur, "Modeling terahertz plasmonic Si FETs with SPICE," *IEEE Trans. Terahertz Sci. Technol.*, 3(5), pp. 545–549, Sep. 2013.

14. W. Knap, D. B. But, N. Dyakonova, D. Coquillat, A. Gutin, O. Klimenko, S. Blin, F. Teppe, M. S. Shur, T. Nagatsuma and S. D. Ganichev, "Recent results on broadband nanotransistor based THz detectors," *in THz and Security Applications*, Springer, Dordrecht, pp. 189–209, Mar. 2014.

15. D. B. But, P. Sai, I. Yahniuk, G. Cywiński, N. Dyakonova, W. Knap, B. Zhang, W. Yan, Z. Li and F. Yang, "Millimetre band detectors based on GaN/AlGaN HEMT," *in 2018 22nd Int. Microw. Radar Conf. (MIKON)*, 2018, pp. 582–584.

16. W. Stillman, C. Donais, S. Rumyantsev, M. Shur, D. Veksler, C. Hobbs, C. Smith, G. Bersuker, W. Taylor and R. Jammy, "Silicon FinFETs as detectors of terahertz and sub-terahertz radiation," *Int. J. High Speed Electron. Syst.*, 20(01), pp. 27–42, Mar. 2011.

17. W. J. Stillman, "Silicon CMOS FETs as Terahertz and Sub-Terahertz Detectors," Ph.D. dissertation, *Dept. ECSE, Rensselaer Polytechnic Institute,* Troy, NY, 2008.

Author Index

A
Agemori, H., 13
Aizin, G., 1

B
But, D. B., 87

C
Che, M., 13
Cywiński, G., 87

D
Doi, R., 13
Dub, M., 87

F
Filipiak, M., 87

H
Hasan, M. M., 119

I
Ivonyak, Y., 87

K
Kacperski, J., 87
Kamiura, Y., 13

Kato, K., 13
Kato, K., 13
Knap, W., 87
Krajewska, A., 87

L
Liu, X., 131

M
Mikalopas, J., 1
Mikami, Y., 13
Mitin, V., 31, 73

O
Otsuji, T., 31, 73

P
Pala, N., 119
Prystawko, P., 87

R
Rumyantsev, S., 87
Ryzhii, M., 31, 73
Ryzhii, V., 31, 73

S
Sai, P., 87
Shur, M., 1, 49, 119, 131